T0297774

Inflammation and the Microcirculation

Integrated Systems Physiology: From Molecule to Function

Editors

D. Neil Granger, *Louisiana State University Health Sciences Center*

Joey P. Granger, *University of Mississippi Medical Center*

Physiology is a scientific discipline devoted to understanding the functions of the body. It addresses function at multiple levels, including molecular, cellular, organ, and system. An appreciation of the processes that occur at each level is necessary to understand function in health and the dysfunction associated with disease. Homeostasis and integration are fundamental principles of physiology that account for the relative constancy of organ processes and bodily function even in the face of substantial environmental changes. This constancy results from integrative, cooperative interactions of chemical and electrical signaling processes within and between cells, organs and systems. This eBook series on the broad field of physiology covers the major organ systems from an integrative perspective that addresses the molecular and cellular processes that contribute to homeostasis. Material on pathophysiology is also included throughout the eBooks. The state-of the-art treatises were produced by leading experts in the field of physiology. Each eBook includes stand-alone information and is intended to be of value to students, scientists, and clinicians in the biomedical sciences. Since physiological concepts are an ever-changing work-in-progress, each contributor will have the opportunity to make periodic updates of the covered material.

Published titles

(for future titles please see the website, www.morganclaypool.com/page/lifesci)

Capillary Fluid Exchange: Regulation, Functions, and Pathology

Joshua Scallan, Virgina H. Huxley, and Ronald J. Korthuis

2010

The Cerebral Circulation

Marilyn J. Cipolla

2009

Hepatic Circulation

W. Wayne Lautt

2009

Platelet-Vessel Wall Interactions in Hemostasis and Thrombosis

Rolando Rumbault and Perumal Thiagarajan

2010

Copyright © 2010 by Morgan & Claypool

All rights reserved. No part of this publication may be reproduced, stored in a retrieval system, or transmitted in any form or by any means—electronic, mechanical, photocopy, recording, or any other except for brief quotations in printed reviews, without the prior permission of the publisher.

Inflammation and the Microcirculation
D. Neil Granger and Elena Senchenkova
www.morganclaypool.com

ISBN: 9781615041657 paperback

ISBN: 9781615041664 ebook

DOI: 10.4199/C00013ED1V01Y201006ISP008

A Publication in the Morgan & Claypool Publishers series

INTEGRATED SYSTEMS PHYSIOLOGY—FROM CELL TO FUNCTION #8

Book #8

Series Editor: D. Neil Granger and Joey Granger, Louisiana State University

Series ISSN

ISSN   2154-560X   print
ISSN   2154-5626   electronic

Inflammation and the Microcirculation

D. Neil Granger
Elena Senchenkova
Louisiana State University

INTEGRATED SYSTEMS PHYSIOLOGY—FROM CELL TO FUNCTION #8

MORGAN&CLAYPOOL LIFE SCIENCES

ABSTRACT

The microcirculation is highly responsive to, and a vital participant in, the inflammatory response. All segments of the microvasculature (arterioles, capillaries, and venules) exhibit characteristic phenotypic changes during inflammation that appear to be directed toward enhancing the delivery of inflammatory cells to the injured/infected tissue, isolating the region from healthy tissue and the systemic circulation, and setting the stage for tissue repair and regeneration. The best characterized responses of the microcirculation to inflammation include impaired vasomotor function, reduced capillary perfusion, adhesion of leukocytes and platelets, activation of the coagulation cascade, and enhanced thrombosis, increased vascular permeability, and an increase in the rate of proliferation of blood and lymphatic vessels. A variety of cells that normally circulate in blood (leukocytes, platelets) or reside within the vessel wall (endothelial cells, pericytes) or in the perivascular space (mast cells, macrophages) are activated in response to inflammation. The activation products and chemical mediators released from these cells act through different well-characterized signaling pathways to induce the phenotypic changes in microvessel function that accompany inflammation. Drugs that target a specific microvascular response to inflammation, such as leukocyte–endothelial cell adhesion or angiogenesis, have shown promise in both the preclinical and clinical studies of inflammatory disease. Future research efforts in this area will likely identify new avenues for therapeutic intervention in inflammation.

KEYWORDS

vasomotor function, leukocyte adhesion, vascular permeability, thrombosis, angiogenesis

Contents

CHAPTER 1

Introduction

Inflammation is typically viewed as a localized protective response to tissue damage and/or microbial invasion, which serves to isolate and destroy the injurious agent and the injured tissue and to prepare the tissue for eventual repair and healing. The survival value of the inflammatory response for both the injured tissue and the animal as a whole is evidenced by the fact that deficiencies of inflammation compromise the host and suggests that inflammation is an important physiological process. In most instances, an inflammatory reaction is short-lived and results in the desired protective response. However, in some cases, excessive and/or prolonged inflammation can lead to extensive tissue damage, organ dysfunction, and mortality. The critical role of inflammation in diseases as diverse as atherosclerosis, diabetes, cancer, reperfusion injury, and Alzheimer's disease [1–5] accounts for the massive research effort that has been directed toward understanding the mechanisms that initiate and regulate the inflammatory response. While the molecular and cellular processes that contribute to inflammation remain poorly understood, advancements in this field of medical research have already facilitated the prevention, control, and cure of different human diseases associated with inflammation. Furthermore, the participation of a variety of signaling pathways, chemical mediators, and cell populations in the inflammatory response provides numerous potential targets for development of novel therapeutics for inflammatory diseases [1–5].

The microcirculation plays a major in the genesis and perpetuation of an inflammatory response. The microvasculature undergoes a variety of functional and structural changes during inflammation. These changes include a diminished capacity of arterioles to dilate, impaired capillary perfusion, the adhesion of leukocytes and platelets in venules, enhanced coagulation and thrombogenesis, an increased vascular permeability to water and plasma proteins, and an eventual increase in the rate of proliferation of blood and lymphatic vessels. The net effects of these responses are to (1) promote the delivery of inflammatory cells to the injured/infected tissue, (2) dilute the injurious/inciting agents in the affected region, (3) isolate this region from healthy tissue and the systemic circulation, and (4) set the stage for the regenerative process. The functional changes are evident during the acute phase of the inflammatory response, i.e., within minutes to hours after initiation of the response. The structural alterations, on the other hand, are manifested during the chronic phase of the inflammatory response, i.e., days after the initial inflammatory insult. These

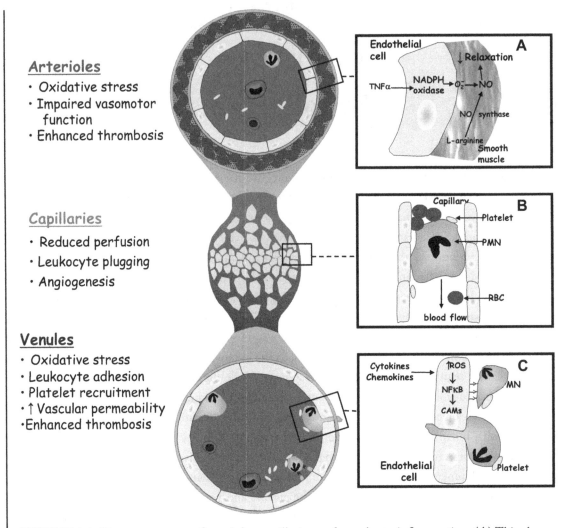

Arterioles
- Oxidative stress
- Impaired vasomotor function
- Enhanced thrombosis

Capillaries
- Reduced perfusion
- Leukocyte plugging
- Angiogenesis

Venules
- Oxidative stress
- Leukocyte adhesion
- Platelet recruitment
- ↑ Vascular permeability
- Enhanced thrombosis

FIGURE 1.1: Diverse responses of arterioles, capillaries, and venules to inflammation. (A) This shows how an inflammatory cytokine can enhance superoxide (O_2^-) formation, inactivate NO, and diminish VSM relaxation in arterioles. (B) This shows how an activated leukocyte can obstruct the capillary lumen and reduce capillary perfusion. (C) This illustrates how activated venular ECs promote the adhesion of leukocytes and platelets to the vessel wall. Modified from Vachharajani and Granger [315].

temporal distinctions between the early and delayed microvascular responses to inflammation are not synonymous with acute vs. chronic inflammatory conditions, where the former is a short-lived self-resolving response that occurs over a period of hours to a few days, while the latter occurs over a period of weeks to years due to a persistent inflammatory stimulus. While the relative contribution of different cellular and humoral mediators may differ between acute and chronic inflammation, the role of the microvasculature in mounting and sustaining the inflammatory response is largely the same under the two conditions [6–10].

All segments of the vascular tree, i.e., arterioles, capillaries, and venules, are affected by, and contribute to, the inflammatory response (Figure 1.1). The activation of endothelial cells (ECs) in each vascular segment by bacterial products and/or locally released mediators allows for a deliberate and coordinated response of the microcirculation to the inflammatory insult (Figure 1.1). Other components of the vascular wall, including vascular smooth muscle (VSM) and pericytes, as well as cells that normally reside in the immediate perivascular region, such as mast cells and macrophages, all contribute to the inflammatory response and serve to induce and/or amplify the events associated with endothelial cell activation. Further amplification and perpetuation of the response results from the recruitment and activation of leukocytes and platelets, which bring their own armamentarium of releasable mediators into the inflamed microvasculature [11–17].

This eBook provides a review of how the microcirculation responds and contributes to inflammation. The underlying mechanisms and physiological consequences of the microvascular responses to inflammation will be addressed from published data derived from both acute and chronic models of inflammation. Relevance to human disease and potential targets for development of novel therapeutics for inflammation are also discussed.

•   •   •   •

CHAPTER 2

Historical Perspectives

The history of inflammation is long and colorful, with descriptions of this process dating back to the ancient Egyptian and Greek cultures. Terms, like *edema*, which are still used to describe inflammation, were introduced by Hippocrates in the 5th century BC. He also regarded inflammation as an early component of the healing process after tissue injury. Aulus Celsus, a Roman writer who lived between 30 BC and 45 AD, described the main four signs of inflammation as redness, warmth, swelling, and pain. We now appreciate that the first three signs are likely attributable to responses of the microvasculature to inflammation. Galen, the physician and surgeon of Roman emperor Marcus Aurelius, is often credited with introducing a fifth sign of inflammation, i.e., loss of function in the affected tissue. Galen also attributed a role for the percolation of blood, which he introduced as one of the four vital humors, in the inflammatory process. While these early concepts about inflammation were largely derived from intuition rather than careful scientific investigation, the controversial observations described by the ancient cultures provided the framework for critical experimentation in the later centuries [6–11].

The invention of the compound microscope by Janssen in the 16th century and the subsequent improvement of its optical resolution by Leeuwenhoek gave rise to the early descriptions of the microcirculation and its responses to inflammation. Eighteenth-century applications of the microscope lead to descriptions of blood flow changes in inflamed tissue and the proposal by Gaubius that inflammation can promote the "disposition to coagulation." In 1794, John Hunter first used the term *angiogenesis* to describe the development of growing blood vessels in healing wounds. Dutrochet (1824) provided the first description of the "sticking" and emigration of white cells (described as "vesicular globules") in blood vessels during acute inflammation, while the first explicit description of leukocyte rolling is attributed to Wagner in 1839. Cohnheim's (1867) classical descriptions of vascular events in thin, transparent tissues in vivo lead him to postulate that "molecular alteration in the vessel walls" underlies the sticking of leukocytes in inflamed microvessels, while their subsequent emigration was attributed to mechanical filtration. Metchnikoff (1893), on the other hand, assigned a more important role for the leukocyte in the emigration process, stating that "the migration is effected by the amoeboid power of the leukocytes." Cohnheim's observations also lead him to speculate that augmented "porousness" of the vessel wall explain the enhanced fluid and protein transudation in inflamed tissue [8,9,18–21].

The 20th century was marked by rapid advancements in understanding the nature and underlying mechanisms of the microvascular responses to inflammation. The development of new in vivo models of inflammation, methods to capture and store images of the microcirculation, and the application of mathematical and engineering approaches to quantify variables such as leukocyte adhesion, vasomotor function, and vascular permeability allowed the field to move forward at a greatly accelerated pace. This period also brought new chemical methods that enabled the discovery of different inflammatory mediators and coagulation factors. With the advent of the electron microscope came the first descriptions of the fine structure of the endothelial cell and other components of the vessel wall. The multifunctional nature of ECs and their critical role in the inflammatory response was born from the explosion of endothelial cell research that began in the 1970s. An outgrowth of this new focus on ECs was the successful effort to isolate and culture ECs from human umbilical veins, which formed the basis for development of in vitro models to study molecular mechanisms underlying endothelium-dependent inflammatory processes such as leukocyte recruitment, thrombosis, increased vascular permeability, and angiogenesis [11,22–24].

In recent years, important additions to the armamentarium of inflammation researchers have come from the fields of molecular biology and immunology. The development of gene-targeted knock-out mice for different inflammatory molecules, such as cytokines, chemokines, and their receptors, as well as leukocyte and endothelial cell adhesion molecules have proven to be immensely useful in the dissection of molecular mechanisms of inflammation in vivo. Immunological approaches (e.g., blocking antibodies, bone marrow chimeras) developed for the mouse, with its exhaustively characterized immune system, have also proven to be powerful tools for the study of inflammation. The large impact of these developments on inflammation research holds promise for further advancements that can result from the continued rapid development of novel technologies and experimental approaches in different areas of biomedical research [8,11,12].

CHAPTER 3

Anatomical Considerations

3.1 MICROVASCULAR UNIT

The architecture of the microcirculation varies among organs, but the fundamental elements of the microvasculature are common to all vascular beds. The microcirculation consists of arterioles, capillaries, and venules, which form a branching, tapered network of approximately circular tubes. Figure 3.1 illustrates the microvascular architecture of skeletal muscle and highlights the dense capillary network that surrounds muscle fibers and the spatial orientation of the capillaries with arterioles and venules. The fundamental interorgan variations in microvascular anatomy relate primarily to branching patterns, vessel densities, and the fine structure of capillaries. Below is a brief

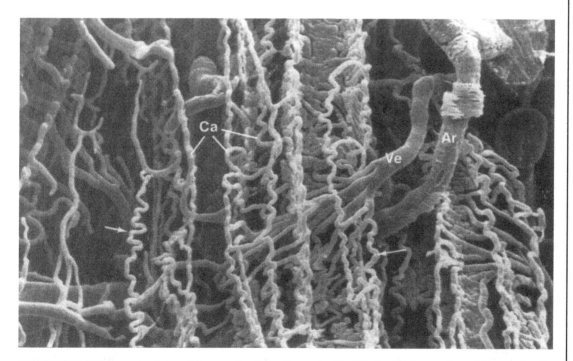

FIGURE 3.1: Microvascular architecture in skeletal muscle. Capillaries (Ca) are oriented longitudinally in the same direction as the muscle fibers they supply. The capillary network is supplied by an arteriole (Ar) and drained by a venule (Ve). Reproduced from Kessel and Karden [27].

description of the three structurally and functionally important elements of the microcirculation, i.e., arterioles, capillaries, and venules [14,25–28].

3.1.1 Arterioles

Each small artery can give rise to several arterioles as its diameter decreases toward the tissue core (or periphery). Arterioles are generally less than 500 μm in diameter, with an external muscular coat that consists of two to four circumferentially arranged smooth muscle cells. As arteriolar diameter decreases, so does the number of smooth muscle layers. The terminal (precapillary) arterioles have an internal diameter of 15 to 20 μm and are surrounded by only one layer of smooth muscle cells. Arterioles, like their parent vessels (large and small arteries), have an inner lining of ECs, and the periphery of the vessel may be invested by fibroblasts and a network of nonmyelinated nerves. Arterioles can undergo active changes in diameter, about twofold to threefold in the smallest vessels and 20% to 40% in the larger arterioles, depending on the initial state of vascular tone. Arterioles account for a majority of total vascular resistance (with capillaries and venules accounting for ≤25%), which imparts a dominant role to the smallest arterioles in the regulation of blood flow under resting conditions. However, this role can be shifted toward larger arterioles as well large conducting arteries under conditions of intense vasodilation [28–30].

3.1.2 Capillaries

Arterioles give rise to capillaries when the internal arteriolar diameter falls below 50 μm, but in the majority of cases, capillaries are derived from terminal arterioles. Capillaries largely consist of a tube (of 4 to 10 μm internal diameter) lined by a single layer of ECs and a thin basement membrane. The ultrastructure and endothelial thickness of capillaries vary considerably between and within organs. Based on the fine structure of their endothelium, capillaries are generally divided into the following categories: (i) fenestrated, (ii) continuous, and (iii) discontinuous. In many tissues, only a fraction (e.g., 20% to 30%) of the capillaries are open to perfusion under resting conditions. The ability of tissues to recruit additional perfused capillaries during periods of stress (e.g., hypoxia) has been attributed to the existence of "precapillary sphincters," which may represent one or two layers of smooth muscle that surround the entrance of a capillary. The capillary network, with its large surface area and an endothelial barrier that is highly permeable to lipid-soluble and small water-soluble molecules, appears well suited for the exchange of gases, nutrients, and water between the bloodstream and tissues [22,26,31,32].

3.1.3 Venules

Capillaries drain into larger vessels that are also devoid of a smooth muscle coat. These postcapillary venules represent the segment of the microvasculature that is most reactive to inflammation and contain intercellular endothelial junctions that can open to allow plasma proteins and circulating cells (leukocytes) to escape from the bloodstream. Indeed, venules represent the major site of trans-vascular protein exchange (vascular permeability to plasma proteins) and leukocyte trafficking (leukocyte–endothelial cell adhesion). The localization of these inflammatory functions in venules is believed to reflect the unique characteristics of ECs in this segment of the microcirculation [22]. Smooth muscle appears on the media of larger venules (muscular venules) that drain the postcapil-lary venules. The organization of the venular network is similar to that of the arterioles except that venules are two to three times wider and are somewhat more numerous (e.g., two to four times in heart) than the arterioles. Furthermore, smooth muscle is more abundant in arterioles than in muscular venules. The passive, distensible nature of the postcapillary and muscular venules accounts for the ability of these microvessels to store and mobilize significant quantities of blood in certain organs [15,22,26,33].

3.2 VESSEL WALL COMPONENTS

ECs and VSM represent the major functional elements of the blood vessel wall that enable arteri-oles, capillaries, and venules to carry out their functions. While the two cell types are clearly capable of functioning independently, there are processes that enable one cell type to influence the other. The phenomenon of endothelium-dependent vasodilation (discussed below) perhaps best exempli-fies how ECs can exert control over the tone of adjacent VSM cells through the production and liberation of vasoactive substances. There are, however, other cells that are either in contact with, or adjacent to, the blood vessel wall that can exert an influence on the activity of ECs and/or VSM cells. Pericytes and mast cells are examples of such auxiliary cells that can exert a profound influence on the function of arterioles, capillaries, and/or venules.

3.2.1 Endothelial Cells

Information derived from both in vivo and in vitro studies of endothelial cell (EC) function has revealed an active role for these cells in mediating an inflammatory response. Characteristically, EC assume an activated state in response to an inflammatory insult. EC activation can be elicited by a variety of chemical (e.g., cytokines) and physical (shear stress) stimuli that are altered during the inflammatory response. For example, transient fluctuations in tissue oxygen levels (pO_2) can result in EC activation. It is unlikely that any single stimulus accounts for inflammation-associated EC activation [23,24,34–37].

Under normal conditions, EC create a highly selective barrier to fluid and solute movement, mediate vasomotor tone, and exhibit surface properties that are both anti-inflammatory, and anti-thrombogenic. The luminal surface of EC is normally lined with a glycocalyx (endothelial surface layer [ESL]), comprised of proteoglycans and glycosaminoglycans, which separates blood cells from the EC membrane by 0.50-1.0 μm and exerts an influence on local hematocrit, flow resistance, oxygen transport, and adhesive interactions between blood cells and the vessel wall. The turnover rate of normal ECs is extremely low (half life of months to years) and only a small percentage (<0.1%) are apoptotic. During inflammation, activated EC undergo a variety of changes that lead to characteristic alterations in microvascular function, including impaired vasomotor function, thrombus formation, leukocyte adhesion and emigration, increased vascular permeability, and angiogenesis. Activated EC also exhibit a reduction in the ESL thickness, and increased rates of apoptosis and detachment from the basement membrane, leading to the appearance of circulating ECs. Both transcription-dependent (delayed) and transcription-independent (rapid) mechanisms underlie the phenotypic changes that are assumed by ECs during inflammation. For example, the mobilization of P-selectin from Weibel–Palade bodies (EC granules) allow for the rapid initiation of leukocyte rolling, while the biosynthesis of E-selectin sustains this recruitment process during inflammation [23,24,36–40].

Reactive metabolites of both oxygen (superoxide) and nitrogen (NO) have been implicated in the endothelial cell response to inflammation. The reactive nitrogen oxide species (RNOS) appear to contribute to the inflammatory process by (1) serving as signaling molecules, (2) activating nuclear transcription factors for both pro-inflammatory and anti-inflammatory proteins, and/or (3) mediating cell necrosis and apoptosis. While NO and superoxide are often ascribed anti-inflammatory and pro-inflammatory roles, respectively, the products of their chemical interaction (RNOS) can yield either phenotype (anti-inflammatory or pro-inflammatory), depending on whether there is net oxidation or nitrosation of specific molecular targets that regulate the inflammatory response (Figure 3.2). Under normal conditions (Figure 3.2A), the balance between NO and reactive oxygen species (ROS) favors an anti-inflammatory phenotype, because NO chemistry predominates as a result of the approximately 1000-fold greater production of NO compared to superoxide in EC. The excess NO yields an anti-inflammatory phenotype through sustained inhibition (related to target-specific nitrosation) of transcription factor activation, a cGMP-mediated transcription-independent signaling. However, in response to inflammatory stimuli (Figure 3.2B), the flux of superoxide relative to NO increases such that ROS-dependent mechanisms predominate. Under these conditions, ROS-mediated transcription-dependent and transcription-independent processes are initiated, the outcome of which not only depends on the relative fluxes of NO and superoxide but also on the specific RNOS formed. The net effect of this imbalance between NO and superoxide is the assumption of a pro-inflammatory phenotype by vascular ECs [41–44].

The fine structure and function of ECs exhibit considerable heterogeneity between organs and between microvascular segments (arterioles, capillaries, venules) within a given organ. Capillaries have been classified as continuous, fenestrated, or discontinuous based on ultrastructural characteristics. This classification accounts for resting differences in the permeability of different vascular beds and it can influence the magnitude of the barrier dysfunction that accompanies inflammation.

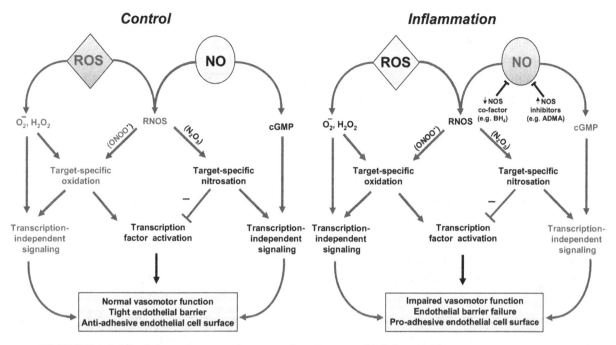

FIGURE 3.2: The balance between the rates of production of ROS and NO is an important determinant of the microvascular responses to inflammation. Under control conditions, the balance between NO and ROS favors normal vasomotor function (arterioles), a tight endothelial barrier, and an anti-adhesive endothelial cell surface because NO chemistry predominates. The excess NO yields this phenotype through sustained inhibition (related to target-specific nitrosation) of transcription factor activation and cGMP-mediated, transcription-independent signaling. During inflammation, the balance between NO and ROS is shifted toward the latter species as a result of a reduction in NO biosynthesis, inactivation of NO by $O2^{\bullet-}$, or both. The production of NO is reduced due to oxidation of cofactors such as tetrahydrobiopterin (BH4) and the accumulation of endogenous NOS inhibitors such as asymmetric dimethylarginine (ADMA). In this instance, the flux of $O2^{\bullet-}$ relative to NO increases such that ROS-dependent mechanisms predominate and NO-dependent mechanisms are rendered inactive. ROS (and possibly RNOS)-mediated transcription-dependent and -independent processes then lead to impaired vasomotor function, endothelial barrier failure, and a pro-adhesive endothelial surface that favors the recruitment of inflammatory cells and thrombosis.

For example, brain capillaries are highly impermeable to even the smallest of solutes and strong agonists/stimuli are needed to disrupt the "blood brain barrier." On the other extreme, tissues perfused by discontinuous capillaries, such as liver, are normally very leaky to the largest plasma proteins, and inflammation elicits minimal or no changes in the restrictive properties of these vessels. A comparison of EC monolayers from macroscopic vs. microscopic blood vessels suggests that the latter are more resistant to agonists that increase vascular permeability. Basal and induced expression of EC adhesion molecules also differs between organs and between arterioles and venules. Most adhesion molecules that contribute to leukocyte recruitment during inflammation are expressed in venules, although some can be induced in arterioles and arteries. The expression of pro-coagulant (e.g., tissue factor) and anti-coagulant (protein C) molecules also has been shown to differ between arterioles and venules. Finally, some specialized ECs (e.g., in high endothelial venules) are known to express unique proteins (e.g., MAdCAM) in a tissue-specific manner. Collectively, these observations indicate that while activated EC typically respond and contribute to inflammation in a predictable manner, significant qualitative and quantitative variations in the response are noted between tissues and within different vessels of the same tissue [22,23,33,35,45,46].

3.2.2 Vascular Smooth Muscle

The changes in microvascular perfusion that accompany inflammation can be attributed to alterations in the tone of smooth muscle surrounding the resistance vessels (arterioles). VSM cells have a highly developed system of contractile and cytoskeletal elements that enable them to contract and relax. The smooth muscle cells respond to a wide variety of vasoactive agents, resulting in either dilation or constriction. Neighboring ECs can exert an influence on VSM tone through the production/release of substances such as NO, prostacyclin, and endothelin, which readily diffuse the short distance between EC and VSM. There is also evidence that ECs and neighboring VSM are electrically coupled to each other in arterioles, allowing electrical signals conducted along EC to be directly transmitted to the surrounding smooth muscle to evoke vasomotor responses [26,29,47,49].

VSM also have the capacity to produce a variety of factors that can contribute to an altered vasomotor function and influence other components of the microvascular response to inflammation. In response to inflammatory cytokines, VSM can produce pro-angiogenic factors such as platelet-derived growth factor (PDGF). These cells are also a rich source of arachidonic acid metabolites such as prostacyclin and PGE_2. Cytokines (IL-1, IL-6) and chemokines (MCP-1) are also generated by activated VSM. Finally, activated VSM can produce ROS and contribute to the oxidative and nitrosative stress that accompanies inflammation [50–52].

3.2.3 Pericytes

These are ameba-shaped, actin-containing cells that are associated abluminally with capillaries and postcapillary venules. These cells extend long, slender processes that are embedded within the basement membrane to directly contact the underlying endothelium. The density of pericytes in a vascular bed varies among tissues and between different-sized vessels. Pericytes are more numerous and have more extensive processes (more contact with endothelium) on venous capillaries and postcapillary venules. Studies on individually cultured and cocultured (with EC) pericytes reveal an ability of these cells to contract and to produce substances that modulate the development and function of ECs (and vice versa). Accordingly, pericytes are considered to play an important role in the regulation of capillary blood flow, capillary growth, and vascular permeability, as well as being precursors to VSM cells [16,52–54].

3.3   PERIVASCULAR AUXILIARY CELLS

Some cells that lie in close proximity to microvessels are known to release chemicals that can alter the function of EC, VSM, and/or pericytes and thereby influence the quality and intensity of the inflammatory response.

3.3.1   Mast Cells

Most notable among these perivascular cells are mast cells, macrophages, and fibroblasts. In many tissues, mast cells are usually found closely apposed to postcapillary venules. These cells are exquisitely sensitive to activation by a variety of stimuli and express receptors for pathogens (TLR-2, TLR-4, and TLR-9), complement (CR2, CR4, and CR5), cytokines (interferon-γ, interleukin-1), and chemokines (CXCR3, CXCR4). Neuropeptides (e.g., substance P), oxygen radicals (e.g., superoxide), lipid mediators (platelet-activating factor, leukotrienes), and bacterial peptides are all capable of eliciting mast cell degranulation. Mast cell-derived modulators of microvascular function include histamine, heparin, vascular endothelial cell growth factor (VEGF), nitric oxide (NO), cytokines (e.g., TNF-α, interleukin-1), chemokines (CCL2, CXCL-9) proteases (e.g., cathepsin G, chymase), and lipid mediators (leukotriene B4, PGE2) [55–58].

3.3.2   Macrophages

Macrophages ("big eaters") are derived from bone marrow as immature monocytes, enter the bloodstream, and emigrate into tissues where they differentiate into resident cells. In most tissues, the macrophages ultimately reside in the perivascular interstitial compartment; however, they may also reside within the vascular space (Kupffer cells in the liver) or in an external compartment (e.g., alveolar macrophages). Activated macrophages have the capacity to generate large fluxes of superoxide (from NADPH oxidase) and NO, which can react to form peroxynitrite and other RNOS. They also produce and liberate cytokines, chemokines, proteases, and angiogenic factors that can influence the intensity of an inflammatory response [59].

3.3.3   Fibroblasts

Fibroblasts ("fiber makers") are connective tissue cells that produce a variety of extracellular matrix components, including collagens, glycosaminoglycans, and fibronectin. While there is phenotypic heterogeneity among fibroblasts from different tissues, these cells typically respond to a variety of stimuli including chemical signals, such as cytokines and growth factors, as well as mechanical forces (stretch) and hypoxia. Activated fibroblasts produce growth factors, including fibroblast growth factor (FGF) and PDGF, matrix metalloproteinases (e.g., MMP-1, MMP-9), cytokines (interleukins and interferons), and anticoagulant proteins (e.g., tissue factor pathway inhibitor, plasminogen activator inhibitor-1). The arsenal of chemicals liberated by fibroblasts enable these cells to contribute to the perpetuation and resolution of an inflammatory response [60,61].

CHAPTER 4

Impaired Vasomotor Responses

4.1   BLOOD FLOW CHANGES

Inflammation is associated with significant alterations in tissue blood flow, the direction and magnitude of which may change as the inflammatory response progresses. In the early phase of an inflammatory response, hyperemia is often noted, which likely accounts for the redness that has been used as a sign of inflamed tissue over the centuries. The hyperemic response probably reflects an initial reaction of arterioles to injury and to the rapid release of vasoactive mediators (histamine, bradykinin, neuropeptides, prostaglandins) produced by mast cells, macrophages, fibroblasts, parenchymal cells, and the vessel wall. The hyperemia has not been attributed to any single chemical mediator. Antihistamines and glucocorticoids have been shown to blunt inflammation-induced vasodilation. It has been proposed that the hyperemic response benefits the host because it helps to rid of (or dilute) the inciting agent and the increased blood flow facilitates the delivery of leukocytes to the site of inflammation [62–65].

ECs may contribute to the early dilatory response of arterioles to inflammation by producing more NO, which relaxes VSM (Figure 4.1). The increased wall shear stress that results secondary to chemical mediator-induced dilation can elicit further vessel relaxation. Selective removal or destruction of the endothelium typically abolishes this response, suggesting the production and/or release of a VSM relaxing factor from EC. Increased shear stress has been shown to elicit a calcium-dependent activation of endothelial cell NO synthase, which oxidizes L-arginine to generate NO and L-citrulline. A role for NO in modulating arteriolar tone is supported by studies employing L-arginine analogs that inhibit NO synthase, as well as L-arginine supplementation to enhance NO production. Although NO has received the most attention as the endothelium-derived factor that mediates flow-dependent vasodilation, prostaglandins, hydrogen peroxide, hydrogen sulfide, and other vasoactive substances have also been implicated in this response [12,29,30,47,48,66,67].

Studies of the blood flow changes that accompany chronic inflammatory diseases suggest that the long-term steady-state response to inflammation is a reduction in tissue blood flow, rather than a hyperemia. In inflammatory bowel diseases (IBDs), for example, intestinal microvascular

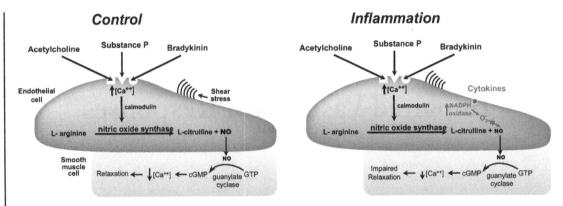

FIGURE 4.1: Endothelium-dependent arteriolar dilation is impaired during inflammation. Under control conditions, ECs respond to receptor-dependent dilators and shear stress by producing NO via calcium-calmodulin-dependent activation of nitric oxide synthase. The NO diffuses into the underlying smooth muscle cell to activate guanylate cyclase, which generates cGMP to reduce intracellular calcium. The net result is smooth muscle relaxation and vessel dilation. During inflammation, cytokines and other mediators activate endothelial NADPH oxidase (and other oxidases) to produce superoxide (O_2^-), which inactivates NO and diminishes the capacity of receptor-dependent agonists and shear stress to elicit dilation in arterioles.

perfusion is increased during the fulminant (active) phase of the disease, while a reduction in flow is detected in chronically inflamed and remodeled intestine. While some animal models of IBD reveal a similar pattern of a transiently increased gut blood, followed by ischemia, other models exhibit relatively small changes in blood flow during the course of the inflammatory response. While the mechanism(s) that underlie the arteriolar dysfunction and decreased blood flow in the chronically inflamed bowel remain poorly defined, reduced responsiveness of arterioles to vasodilators and an enhanced sensitivity to vasoconstrictors have been offered as explanations [62,63,65,68–72].

4.2   ENDOTHELIUM-DEPENDENT VASODILATION

Recent evidence implicates endothelial cell activation and a loss of NO-dependent dilation in the diminished blood flow that accompanies inflammation. While normal arterioles from human intestinal submucosa dilate in a dose-dependent and endothelium-dependent manner to acetylcholine, arterioles in the chronically inflamed intestine show a significantly reduced dilation response to acetylcholine. The inflamed arterioles exhibit an enhanced oxidative stress compared to control arterioles and treatment of arterioles from inflamed human intestine with a superoxide dismutase (SOD) mimetic restores the acetylcholine-induced vasodilatory response to a normal level (Figure 4.2A). The arterioles were also shown to be heavily dependent on cyclo-oxygenase (COX)-

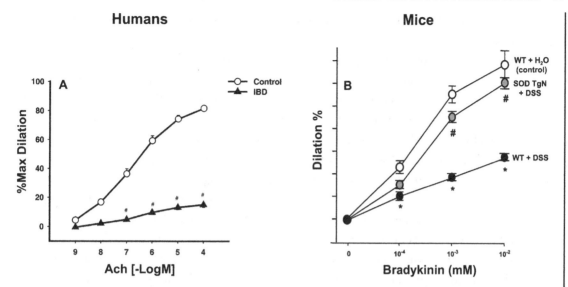

FIGURE 4.2: Impaired endothelium-dependent vasodilation during intestinal inflammation in (A) humans with IBD and (B) in mice with dextran sodium sulfate (DSS)-induced colitis. Control intestinal arterioles exhibit a brisk dilatory response to (A) acetylcholine and (B) bradykinin. Gut inflammation is associated with a diminished capacity of the arterioles to dilate the same agonists. Mice that genetically overexpress SOD TgN are protected (relative to their WT counterparts) against the DSS-colitis-induced inhibition of endothelium-dependent vasodilation. Data derived from (A) Hatoum et al. [73] and (B) Mori et al. [68]. Reproduced from Colonic blood flow responses in experimental colitis: time course and underlying mechanisms, Mori M, Stokes KY, Vowinkel T, Watanabe N, Elrod JW, Harris NR, Lefer DJ, Hibi T, Granger DN, with permission of the American Physiological Society.

derived vasodilator products to maintain basal vascular tone and dilator capacity to acetylcholine. Overall, these findings in human IBD are consistent with a role for ROS-dependent impairment of the vasodilatory capacity of arterioles in the impaired tissue perfusion and oxygenation that occurs in chronically inflamed tissue [73].

The changes in blood flow and impaired arteriolar reactivity to endothelium-dependent vasodilators described in human IBD have been recapitulated in a widely employed mouse model of experimental colitis: dextran sodium sulfate (DSS). Both a significant reduction in blood flow and an impaired reactivity to bradykinin (an endothelium-dependent vasodilator) was noted in the smallest arterioles of wild type (WT) mice with DSS colitis (Figure 4.2B). However, the inflammation-induced vasomotor dysfunction was not evident in mutant mice that overexpress Cu,Zn-SOD and in mice that are genetically deficient in the NAD(P)H oxidase subunit gp91(phox). These findings support the involvement of superoxide in the impaired vasomotor response to inflammation and suggest that NAD(P)H oxidase is a major source of the ROS that mediates this vascular response [68].

A defective endothelium-dependent vasodilatory response has been described in many chronic pathologic conditions that include a significant inflammatory component such as atherosclerosis, diabetes, obesity, and hypertension. Similarly, both localized (ischemia–reperfusion [I/R]) and systemic (sepsis) acute inflammatory responses also manifest the same vasomotor impairment. A role for ROS-mediated inactivation of NO has been implicated in the impaired endothelium-dependent dilation associated with all of these conditions. However, other inflammatory factors, including cytokines, C-reactive protein (CRP), and circulating oxidized low-density lipoprotein (oxLDL), have also been implicated in this response. Anti-inflammatory therapies, such as aspirin, statins, cytokine-directed antibodies, have been shown to improve endothelium-dependent vasodilation. While the effects of short-term exposure of resistance vessels to cytokines yield variable changes in tone, long-term exposure appears to enhance vessel responsiveness to constrictors (e.g., endothelin) and to impair endothelium-dependent vasodilation. Of the cytokines studied to date, TNF-α and IL-1β have received the most attention for their ability to impair endothelial cell and vasomotor functions. TNF-α may well exert its effects by enhancing superoxide formation via NADPH oxidase; however, there is also evidence that the cytokine impairs the stability of eNOS mRNA [66,67,74–78].

Leukocytes can also exert a modulating influence on endothelium-dependent vascular reactivity. Various tissues exposed to I/R exhibit an impaired reactivity of small arteries and arterioles to acetylcholine and other endothelium-dependent vasodilators. It has been shown that animals receiving monoclonal antibodies against leukocyte adhesion molecules or that are genetically deficient in the same adhesion glycoproteins show an improved dilator response to acetylcholine after I/R, suggesting that leukocyte–endothelial cell adhesion contributes to the impaired vasomotor function. These observations, coupled with the protective actions of SOD treatment in the same models, suggest that adherent leukocytes, rather than ECs, may be the major source of superoxide that inactivates NO in this model of acute inflammation [51,79].

CHAPTER 5

Capillary Perfusion

In healthy tissue, the rate and extent of capillary perfusion are governed by the tone of upstream arterioles and precapillary sphincters. The number of capillaries open to perfusion (and precapillary sphincter tone) is usually determined by the metabolic demands of the tissue and its need for oxygen, which is consistent with the classic concept that alterations in functional capillary density allow local modulation of O_2 exchange area and capillary-to-cell diffusion distances. This finely tuned process appears to be compromised during acute and chronic inflammatory states. An extreme form of this effect of inflammation is seen in sepsis, which is associated with a loss of functional capillary density (number of perfused capillaries), impaired regulation of oxygen delivery, and a rapid onset of tissue hypoxia despite adequate delivery of oxygen to the organ. The systemic inflammatory response that accompanies sepsis also leads to disseminated intravascular coagulation and the formation of microclots that obstruct capillaries. Inflammation-mediated events occurring in arterioles, capillaries, and venules can all result in impaired capillary perfusion (Figure 5.1) [25,31,80].

Intravital microscopic observations of leukocyte movement through normal capillaries have revealed a slow passage that can result in a "pile-up" of red cells upstream in the same vessel as well as an intermittent (stop–go) flow in the vessels. This steric hindrance related malperfusion of capillaries by slow-moving leukocytes is exacerbated by inflammation. Reductions in local perfusion pressure, systemic, or local leukocyte activation, as well as the contraction of pericytes that surround capillary endothelium, could explain the impaired capillary perfusion that accompanies inflammation. While the initial trapping of leukocytes in capillaries may be mechanical (steric hindrance), the subsequent of activation of these slow-moving leukocytes by mediators released from the inflamed tissue induce the rapid expression of adhesion molecules (e.g., CD11b/CD18) on their surface and also make the cell less deformable. The rigid phenotype assumed by the activated leukocyte is a result of the polymerization of actin within the cell. The rigid leukocytes that escape the inflamed tissue also have the potential to lodge in capillaries of a downstream vascular bed, such as liver and lung. The low pressure and long narrow capillaries of the lung make this vascular bed an ideal filter for the entrapment of activated leukocytes that gain access to the systemic circulation [81–84].

Activated neutrophils have been shown to leave the inflamed gut and accumulate in liver sinusoids, causing a reduction in the number of perfused sinusoids (capillaries) and tissue hypoxia

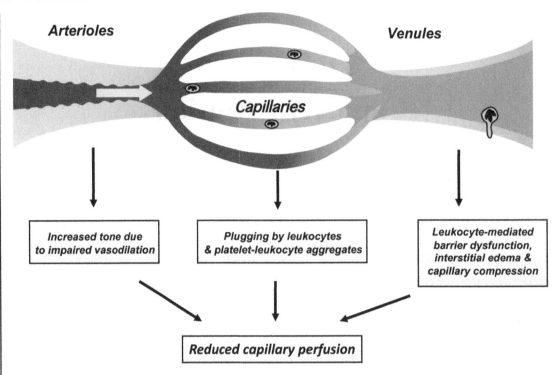

FIGURE 5.1: Roles of arterioles, capillaries, and venules in the reduced capillary perfusion that accompanies an inflammatory response. The increased arteriolar tone, leukocyte–capillary plugging, and leukocyte-mediated fluid filtration from venules all tend to reduce capillary perfusion.

in the liver of WT mice. The leukostasis and tissue hypoxia are blunted in mice that are deficient in either ICAM-1, P-selectin, or CD11/CD18, which suggests that simple steric hindrance is less important than leukocyte–endothelial cell adhesion in mediating this response. Other studies have also implicated the intravascular Kuppfer cell (macrophages) in this response, as well as a role for T-lymphocytes. Neutrophil accumulation, loss of perfused sinusoids, or tissue hypoxia is not observed in immunodeficient SCID mice; however, the adoptive transfer of WT T-cells into the SCID restores the inflammatory responses. The T-lymphocyte derived cytokine IFN-γ has also implicated in the loss of perfused liver capillaries caused by gut-derived neutrophils [85–88].

The reduced capillary perfusion caused by obstruction from activated leukocytes and/or from aggregates of leukocytes and platelets has also been implicated in the "capillary no-reflow" phenomenon that has been described in postischemic tissues. The propensity for capillary plugging with leukocytes in ischemic tissues has been attributed to a number of events, including increased stiffness of activated leukocytes, endothelial cell swelling, leukocyte–endothelial cell adhesion, and low driving pressures for leukocyte movement along the capillaries. It is likely that the relative contribution of each of these factors to capillary no-reflow varies between organs. Thus, organs (e.g., heart) that are perfused by capillaries with small internal diameters will be more sensitive to leukocyte plugging during and following periods of hypotension. Other tissues (e.g., skin) that have large arterial–venous anastomoses, which shunt leukocytes past the capillary bed, will be less prone to leukocyte capillary plugging. The physical restriction or trapping of leukocytes within capillaries has been implicated as a major contributor to the leukocyte accumulation observed in postischemic myocardium, liver, brain, kidney, and skeletal muscle. The strong correlation between the percentage of capillaries exhibiting no-reflow and the percentage of capillaries that contain granulocytes in postischemic tissues suggests that leukocyte trapping likely accounts for capillary no-reflow. Additional supportive evidence is provided by reports that demonstrate virtual elimination of capillary no-reflow in animals that are either rendered neutropenic or receive antibodies that interfere with leukocyte–endothelial cell adhesion in postcapillary venules. Since the level of adhesion molecule expression on capillary endothelium is quite low (compared to venular endothelium), the ability of adhesion molecule-directed antibodies to attenuate capillary no-reflow has been attributed to an action on the downstream postcapillary venules. Adherent leukocytes in postcapillary venules appear to promote leukostasis in upstream capillaries by enhancing fluid and protein filtration across venular endothelium. The resulting interstitial edema raises interstitial fluid pressure to a level sufficient to occlude the capillary lumen and thereby facilitate the entrapment of leukocytes [4,28,82,85,89–91].

· · · ·

CHAPTER 6

Angiogenesis

While capillary growth and proliferation are rarely observed in normal adult tissues except during wound healing and cyclical events in the female reproductive cycle (ovulation, menstruation), with appropriate stimuli, the process of angiogenesis (development of new blood vessels from an existing vascular network) can be initiated (Figure 6.1). ECs exposed to such stimuli first detach from each other through alterations in adherens junction complexes, and metalloproteinases are then released to degrade the underlying basement membrane and surrounding structural elements. Hence, the initiation of angiogenesis is often associated with an increased capillary permeability that serves to enrich the adjacent interstitial compartment with plasma components. Upon destabilization of the endothelial cell monolayer, the cells then migrate (haptotaxis) toward the angiogenic stimulus within the extravascular space via integrin ($a_v\beta_3$, $a_v\beta_5$)-mediated adhesion to matrix proteins, with a concomitant proliferation of the ECs lining the vessel wall to replace the previously migrated cells. The migrating and proliferating ECs form cord-like structures in target tissues that later canalize to form functional vessels, which are further stabilized by surrounding pericytes. Tight cell–cell adhesion results from the expression and function of different adhesion molecules such as PECAM-1 and VE cadherin [92–94].

6.1   RELEVANCE TO INFLAMMATION

While angiogenesis is normally a tightly controlled process that rarely occurs in the adult organism, a number of pathological conditions are known to be associated with aberrant angiogenesis. These conditions include cancer, diabetic retinopathy, ischemic cardiovascular diseases (e.g., stroke), and chronic inflammatory diseases (e.g., IBD). Although the link between angiogenesis and inflammation has received much attention in recent years, there has long been evidence suggesting that these are two closely related processes. These include the appearance of newly formed blood vessels in granulation tissue, and the dual functionality of angiogenic factors, i.e., they exhibit both pro-inflammatory and pro-angiogenic effects. It should be emphasized that while inflammation and angiogenesis are capable of potentiating each other, these processes are distinct and separable. Nonetheless, there is growing evidence that the angiogenesis that accompanies chronic inflammation tends to prolong and intensify the inflammatory response. This contention is supported by

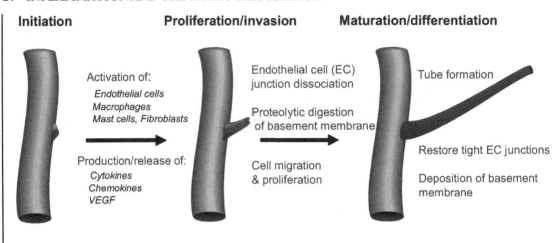

FIGURE 6.1: Events associated with inflammation-induced angiogenesis. During inflammation, angiogenesis is initiated by the activation of different cell populations, which release a variety of angiogenic factors. The next stage (proliferation/invasion) involves changes in the vessel wall that allows for the migration and proliferation of ECs. The final stage (maturation/differentiation) includes tube formation and restoration of a normal vessel wall. The rate of angiogenesis is determined by the balance of angiogenic and angiostatic factors.

reports describing a worsening of disease activity, tissue injury, and colonic inflammation in experimental IBD by administration or genetic overexpression of VEGF-A while treatment of colitic mice with anti-angiogenic agents or genetic overexpression of soluble VEGFR-1 had the opposite effect. These findings have led to the proposed use of anti-angiogenesis drugs in the treatment of IBD [92,95–104].

6.2 MEDIATORS OF THE ANGIOGENIC RESPONSE

There is an abundance of factors produced by mammalian tissues that are capable of inhibiting or promoting blood vessel proliferation (Figure 6.1). Hence, the balance between these angiogenic and angiostatic factors determines the existence and rate of blood vessel proliferation in a tissue.

In inflammation, this balance is clearly tipped in favor of angiogenesis. This response results, in part, because an inflammatory locus is often hypoxic and hypoxia is an important pro-angiogenic signal that activates the hypoxia-inducible factor signaling pathway, which elicits the transcription-dependent production of VEGF and FGF. Inflammation is also associated with the recruitment of circulating leukocytes and platelets, and the activation of resident macrophages, mast cells, and fibroblasts, all of which are capable to producing large quantities of pro-angiogenic factors, including VEGF and cytokines [93,94,97].

6.2.1   Vascular Endothelial Cell Growth Factor

The VEGF family and its receptors (VEGFR-1 and VEGFR-2) have long been implicated as a central figure in the regulation of angiogenesis. VEGF-A directly stimulates EC proliferation by engaging with the VEGFR-2 to activate tyrosine kinase and initiate the sprouting of new vessels from existing microvessels. Sprouting requires the destabilization of existing microvessels, which includes pericyte dropout, diminished cell–cell adhesion, and dissolution of the basement membrane. This process begins with an endothelial ("tip") cell leaving the endothelial monolayer, penetrating the basement membrane, and invading the adjacent interstitial compartment. The tip cell assumes a distinct phenotype that enables it to produce lamellopodia that extend ahead of the cell to "taste" the environment and determine the appropriate migratory direction. Following the tip cell are migratory/proliferative cells ("trunk" or "stalk" cells), which allow for extension of the sprouting vessel and lumen formation. VEGF appears to mediate several steps in this process of sprout formation, including pericyte dropout, induction of tip cell migration, and the formation of lamellipodia, and to provide a substrate for tip cell chemotaxis in the interstitium. VEGF also diminishes the intensity of the endothelial cell–cell interactions in the angiogenic sprout by promoting the phosphorylation and internalization of VE-cadherin. The weakened cell–cell adhesion enables tip cells to depart from the endothelial lining and also accounts for the increased vascular permeability that is characteristic of proliferating blood vessels. While most of the angiogenic effects of VEGF are mediated through VEGFR-2, the hyperpermeability associated with this process requires the activation of both VEGFR-1 and VEGFR-2. Finally, it is noteworthy, in view of recent evidence implicating bone marrow-derived cells in the formation of new blood vessels in chronic inflammatory diseases, that VEGF may also contribute to the angiogenic process by mobilizing endothelial progenitor cells and other myeloid cells to the site of angiogenesis [94,100,104–107].

6.2.2   Cytokines and Chemokines

Infiltrating and resident inflammatory cells, ECs, and VSM all have the capacity to generate large amounts of cytokines and chemokines. Some of these substances exert significant pro-angiogenic or anti-angiogenic properties that may influence the intensity of the angiogenic response elicited during inflammation. Some of the effects on angiogenesis that are attributable to cytokines relate to their ability to prime ECs for the subsequent actions of VEGF. For example, TNF-α has been implicated in the priming of endothelial "tip cells" for migration induced by VEGF, an action that relates to the initial inhibition of VEGFR signaling induction of the "tip cell" phenotype. A more direct and potent action on angiogenesis is noted for the CXC family of chemokines with the glutamic acid–leucine–arginine (ELR) motif immediately proximal to the CXC sequence. The ELR-positive CXC chemokines, including CXCL-2 and IL-8/CXCL8, promote angiogenesis, while ELR-negative CXC chemokines, such as CXCL-9 and CXCL-10, are angiostatic. The primary receptor for chemokine-mediated angiogenesis is CXCR2, which is expressed, along with CXCR1, on ECs. This is supported by the observation that CXCR2 knockout mice and WT mice treated with CXCR2 neutralizing antibodies exhibit a blunted CXC chemokine-mediated angiogenic response. Members of the CC chemokine family that are also pro-angiogenic include, CCL2, CCL11, and CCL16. The engagement of CCL2 with its receptor (CCR2) on EC elicits chemotaxis and tube formation in vitro, and the chemokine has been shown to promote angiogenesis in vivo. CCL2 as well as IL-8 can mediate the homing of circulating endothelial progenitor cells to sites of inflammation. Finally, CCL2-induced angiogenesis has been associated with the induction of VEGF-A gene expression, suggesting that the chemokine works in concert with VEGF to promote angiogenesis during inflammation [94,96,108–110].

6.2.3 Reactive Oxygen and Nitrogen Species

Angiogenesis involves the activation of a variety of signaling pathways. While VEGF and cytokines/chemokines can stimulate different components in the angiogenesis process through different signaling pathways, the balance between NO and ROS production appears to be an important modulator of the angiogenic response to inflammation. VEGF and many cytokines are known to activate NADPH oxidase in vascular EC, likely as a result of the activation and translocation of the small GTPase Rac1 into the plasma membrane. There is also evidence that VEGF elicits the activation of eNOS, which is largely localized in caveolae, via a PI3K/Akt-dependent mechanism. NADPH oxidase-dependent ROS and eNOS-dependent signaling appear to influence different components of new vessel formation, including endothelial junction destabilization, MMP activation, EC migration, and tube formation. The importance of eNOS localization to EC calveolae is evidenced by the observation that caveolae-deficient ECs cannot migrate. The relevance of this observation to inflammatory disease is highlighted by studies demonstrating that caveolin (Cav-1)-deficient mice or WT mice treated with a Cav-1 inhibitory peptide exhibit a blunted angiogenic response to colonic inflammation. Mice that overexpress Cav-1 only in the endothelium also respond to inflammation in a manner consistent with endothelial Cav-1 as an important mediator of angiogenesis in experimental colitis. It remains unclear, however, whether the contribution of caveolae to inflammation-induced angiogenesis is related entirely or in part to localized production of NO and/or ROS [111–114].

6.3 LYMPHANGIOGENESIS

The enhanced vascular proliferation that accompanies inflammation is not limited to blood vessels. There is growing evidence for an increased abundance of lymphatic capillaries in inflamed tissues. Both acute inflammatory stimuli (e.g., bacterial infection) or chronic inflammatory diseases (e.g., human IBD) are associated with lymphangiogenesis. It has been proposed that failure of lymphatic pumping caused by a direct action of inflammatory mediators, such as prostaglandins (PGE2 PGI2) or cytokines, may lead to lymphatic insufficiency and a compensatory proliferative response. Alternatively, the lymphangiogenesis may result from a stimulatory action (mediated via NFkB) of cytokines such as TNF-α on different cells that produce and release of VEGF-C, a potent stimulant of lymphatic proliferation. Activated macrophages and granulocytes, which produce large amounts of VEGF-C and VEGF-D, have also been implicated in the lymphangiogenic response to inflammation. A role for VEGF-C in this response is supported by studies demonstrating that VEGFR-3-Ig ligand trap, which blocks VEGF-C and VEGF-D, suppresses the inflammation-induced lymphangiogenesis. Blocking integrin $\alpha5\beta1$ signaling with small-molecule inhibitors also inhibits the lymph vessel proliferation associated with inflammation, presumably by directly inhibiting lymphatic endothelial cell proliferation and migration. The participation of soluble factors in this response is evidenced by the fact that lymphatic vessel proliferation also occurs in lymph nodes that perfused by lymph that drains the inflamed tissue. While the pathophysiological consequences of inflammation-induced lymphangiogenesis remain unclear, it has been suggested that the response helps to rid the inflamed tissue of edema fluid and it may facilitate the clearance of immune cells (macrophages, dendritic cells) from the tissue. This contention is supported by studies demonstrating that inhibition of inflammation-associated lymphangiogenesis increases the severity of the inflammatory cell accumulation and interstitial edema [115–119].

CHAPTER 7

Leukocyte–Endothelial Cell Adhesion

The adhesion of leukocytes to vascular endothelium is a hallmark of the inflammatory process. This recruitment process and the requirement for (and participation of) specific adhesion glycoproteins in the binding of leukocytes to ECs have been elegantly demonstrated using a variety of experimental approaches. Direct visualization of the inflamed microvasculature has revealed that as leukocytes exit capillaries, hemodynamic forces give rise to an outward radial movement of leukocytes toward the venular endothelium. This margination process is generally attributed to red blood cells (which normally pile up behind the larger leukocytes in capillaries) that overtake the leukocytes and tend to push them toward the venular wall. The initial adhesive interactions between the leukocytes and venular endothelium are tethering (capture) and rolling. These low-affinity (weak) interactions are subsequently strengthened as a result of leukocyte activation (mediated by chemokine-dependent and chemokine-independent mechanisms). Consequently, the leukocytes attach to the endothelium and remain stationary. The leukocytes are then able to migrate into the interstitium through spaces between adjacent ECs. These interactions are initiated by a variety of chemical mediators that are elaborated from inflamed tissue and the entire process of leukocyte–endothelial cell adhesion is regulated by the sequential activation of different families of adhesion molecules that are expressed on the surface of leukocytes and ECs. The current paradigm for leukocyte (neutrophil) recruitment in the inflamed microvasculature is summarized in Figure 7.1. The adhesive determinants are known to vary between vascular beds and between different leukocyte populations [17,37,46,120–123].

7.1   ADHESION MOLECULES

Table 7.1 summarizes some of the adhesion molecules that are expressed on the surface of leukocytes and their respective counter-receptors on ECs. Lectin-like adhesion glycoproteins, called the *selectins*, mediate leukocyte rolling, while the firm adhesion and subsequent transendothelial migration of leukocytes are mediated by the interaction of integrins (CD11/CD18, VLA-4) on leukocytes with immunoglobulin-like adhesion molecules on ECs (e.g., ICAM-1, VCAM-1). The expression of P-selectin, E-selectin ICAM-1, and VCAM-1 on venular EC are temporally coordinated to ensure that the processes of leukocyte rolling and firm adhesion/emigration can occur for

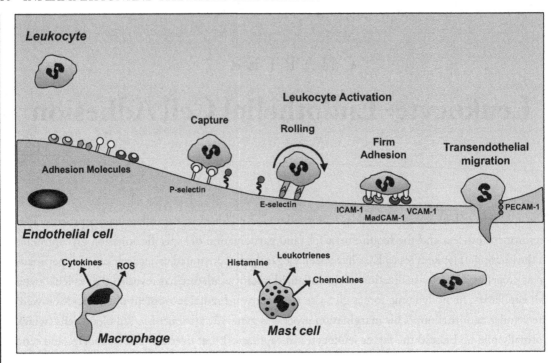

FIGURE 7.1: Multistep adhesion cascade and the major molecular contributors to leukocyte recruitment during an inflammatory response. Following endothelial cell activation and the increased expression of P- and E-selectins, low-affinity adhesive interactions (capture and rolling) are elicited that subsequently lead to leukocyte activation, followed by firm adhesion and transendothelial migration. The ligands for the adhesion receptors depicted here are summarized in Table 7.1.

several hours after the initiation of an inflammatory response. The importance of these endothelial cell adhesion molecules and their counter-receptors on leukocyte recruitment in different animal models of inflammation has been demonstrated using either adhesion molecule-specific blocking monoclonal antibodies (mAbs) or mice that are genetically deficient in one or more adhesion molecules [37,79,86,122–127].

7.1.1   Intraorgan Heterogeneity of Adhesion

The sequential, coordinated recruitment of leukocytes into the inflamed microvasculature is largely confined to postcapillary venules, with leukocyte–endothelial cell adhesion rarely seen in arterioles. For example, in the rat mesenteric microcirculation, 39% of all leukocytes passing through venules are rolling, while only 0.6% of leukocytes roll in the upstream arterioles. The basis for the preferential binding of leukocytes to venular ECs during inflammation appears to relate to the

TABLE 7.1: Leukocyte adhesion receptors and their ligands on activated ECs.

LEUKOCYTE ADHESION RECEPTOR	ENDOTHELIAL LIGAND	FUNCTION(S)
PSGL-1	P-selectin	Capture
		Rolling
L-selectin	P-selectin	Capture
	Peripheral node addressin (PNAd)	Rolling
	E-selectin	
	MadCAM-1	
$\alpha_4\beta_7$ (unactivated)	MadCAM-1	Rolling
$\alpha_4\beta_7$ (activated)	VCAM-1/MAdCAM-1	Firm adhesion
$\alpha_4\beta_1$ (unactivated)	VCAM-1	Rolling
$\alpha_4\beta_1$ (activated)	VCAM-1	Firm adhesion
CD11a/CD18 (LFA-1)	ICAM-1, ICAM-2	Firm adhesion
		Emigration
CD11b/CD18 (Mac-1)	ICAM-1	Firm adhesion
		Emigration
PECAM-1	PECAM-1	Emigration

higher expression of endothelial cell adhesion molecules (CAM) in venules. Immunohistochemical analyses of CAM expression in the microcirculation have yielded results that consistently show a preferential expression of endothelial CAMs in postcapillary venules. Endothelial cell adhesion molecules such as intercellular adhesion molecule-1 (ICAM-1), vascular cell adhesion molecule-1 (VCAM-1), E-selectin, and P-selectin can be detected on the surface of activated ECs in arterioles and occasionally capillaries; however, the density of these adhesion molecules is far greater on venular endothelium. Quantitative information bearing on this issue has been generated using laser confocal microscopy to quantify the constitutive expression of ICAM-1 in arterioles, capillaries, and venules of rat mesentery, using an FITC-labeled anti-rat monoclonal antibody. Mesenteric venules exhibited a 10-fold higher density of ICAM-1 than in the upstream arterioles and capillaries. This approach also revealed heterogeneity of ICAM-1 expression within different sized venules. Venules with diameters of 25 μm appear to exhibit the greatest density of ICAM-1 on ECs, while 15 and 35–40 μm diameter venules exhibit the lowest constitutive expression. This venule size-dependent distribution of ICAM-1 is consistent with functional evidence demonstrating that 25- to 30-μm diameter venules sustain the most intense leukocyte adhesion responses to pro-inflammatory mediators. It remains unclear whether other endothelial CAMs exhibit a similar size-dependent distribution within postcapillary venules [11,33,46,120–123,128–130].

7.1.2 Selectins

Selectins are lectin-like adhesion glycoproteins that mediate leukocyte rolling, which serves to sufficiently reduce the velocity of leukocyte movement along endothelial cell to allow for firm adhesion. P-selectin can be expressed by both EC and platelets, while E-selectin is only expressed by EC. Some vascular beds, such as intestine, exhibit significant constitutive expression of P-selectin, while only skin microvessels exhibit basal expression of E-selectin. P-selectin is normally stored as a preformed pool in EC granules (Weibel–Palade bodies), from which it can be rapidly mobilized to the cell surface by histamine, ROS, and leukotrienes. A slower, more prolonged (transcription-dependent) expression of P-selectin can be demonstrated within 4 hours after exposure to cytokines such as TNF-α. E-selectin, which does not exist in a preformed pool, is entirely under transcriptional regulation and requires up to 3 hours to achieve peak expression. The kinetics and magnitude of expression of P- and E-selectin varies between tissues, with the largest increments in both P- and E-selectin expression noted in the lung, small intestine, and heart after endotoxin challenge, while the brain and skeletal muscle exhibit the smallest responses. Histamine-induced P-selectin expression in all vascular beds is inhibited by a histamine-H_1 (but not an H_2)-receptor antagonist, indicating that histamine engagement of H_1-receptors on ECs results in the mobilization of preformed P-selectin to the endothelial cell surface. The time-related changes in endothelial P-selectin expression after histamine treatment, as well as the temporal responses of P- and E-selectin to endotoxin, are shown for the intestinal microvasculature in Figure 7.2 [120,121,131–140].

Leukocytes also express an adhesion molecule belonging to the selectin family (L-selectin) as well as counter-receptors for EC selectins (PSGL-1). L-selectin is constitutively expressed on most leukocytes, where it is situated on the tips of microvillus cell surface protrusions at a high density. L-selectin mediates leukocyte rolling by interacting with P- and E-selectins expressed on EC. Upon activation of the leukocyte, L-selectin is rapidly shed from the cell surface via a protease-

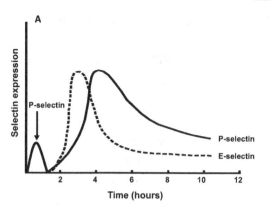
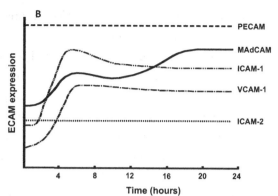

FIGURE 7.2: Kinetics of expression of endothelial cell adhesion molecules that mediate either the (A) rolling or (B) firm adhesion/emigration of leukocytes in postcapillary venules of mouse intestine. The rapid, short-lived expression of P-selectin in (A) was induced by the engagement of histamine with H1 receptors on ECs. The slow, prolonged up-regulation of P- and E-selectin was elicited by endotoxin. The responses in (B) were induced by TNF-α. Data derived from Eppihimer et al. [132], Eppihimer and Granger [133], and Henninger et al. [145]. Reproduced from Heterogeneity of expression of E- and P-selectins in vivo, Eppihimer MJ B Wolitzky, DC Anderson, MA Labow, DN Granger, with permission of Springer-Verlag New York inc.

dependent mechanism. P-selectin glycoprotein ligand-1 (PSGL-1) is the most important selectin ligand expressed on leukocytes. PSGL-1 expressed on neutrophils and monocytes is constitutively active and can bind to P- and E-selectins, as well as L-selectin. On lymphocytes, PSGL-1 must be enzymatically glycosylated for selectin binding. Recent evidence indicates that EC can also express low levels of PSGL-1 [120,121,140–143].

7.1.3   Endothelial Cells Immunoglobulin-Like Adhesion Molecules

ICAM-1, ICAM-2, VCAM-1, and PECAM-1 all belong to a family of immunoglobulin-like molecules that are expressed on the surface of EC. These molecules engage with leukocyte counter-receptors to mediate firm adhesion and/or transendothelial migration. ICAM-1 and ICAM-2 exhibit significant constitutive expression on EC in most vascular beds. When normalized for inter-organ differences in EC surface area, the lung exhibits the highest density of ICAM-1, followed closely by the small intestine. Even though other organs (e.g., brain) express much lower levels of ICAM-1, the density of ICAM-1 in all organs is 100 to 1000 times greater than that observed for P- or E-selectin in the same tissues. The observation that constitutive ICAM-1 expression is significantly reduced in microvessels of the GI tract and skin of germ-free mice compared to their conventional counterparts suggests that indigenous gastrointestinal microflora are responsible for a significant proportion of the basal ICAM-1 expression that is detected in both intestinal and extra-intestinal tissues [120,121,140,144–146].

The constitutive expression of ICAM-2 and PECAM-1 on EC in most vascular beds is high, and these adhesion glycoproteins are not up-regulated in response to cytokine challenge. ICAM-1 and VCAM-1, on the other hand, respond to cytokine or endotoxin challenge with a time- and dose-dependent increase in EC expression that is transcription-dependent. The oxidant-sensitive transcription factors, NFkB and AP-1, play a major role in linking EC activation with a variety of different stimuli to increased ICAM-1 and VCAM-1 expression in inflamed microvessels. The transient surge in circulating soluble ICAM-1 (sICAM-1) concentration that precedes the increased surface expression of ICAM-1 on EC suggests that there is a rapid, massive shedding of membrane-bound ICAM-1 from EC throughout the inflamed microvasculature and the eventual clearance of shed protein from the circulation. The shedding of membrane-bound adhesion glyco-protein from activated EC accounts for the use of soluble circulating adhesion molecules as a surro-gate marker for EC activation and the severity of inflammatory response [120,121,145,147,148].

7.1.4 Leukocyte Integrins

Integrins are glycoprotein complexes consisting of α- and β-subunits. Once integrin function is activated, the glycoprotein can mediate a strong adhesive interaction (firma adhesion) with counter-receptors on EC. For example, VLA-4 is an α4/β1 integrin that enables monocytes and lymphocytes to bind to VCAM on EC. The β2 (CD11/CD18) integrins play a more important role in mediating the firm adhesion of neutrophils. The CD11/CD18 complex is composed of three structurally and functionally related glycoprotein heterodimers comprised of a distinct α-subunit (CD11a, CD11b, CD11c) that is noncovalently bound to a common β-subunit (CD18). Both CD11a and CD11b are constitutively expressed on the surface of most leukocytes. Most of the CD11b/CD18 and CD11c/CD18 glycoproteins are stored in granules and can be rapidly (within minutes) mobilized to the surface of leukocytes. Following leukocyte activation, some heterodimers (e.g., CD11b/CD18) are mobilized to the cell surface while the constitutively expressed CD11a/CD18 (which is not stored in granules) heterodimer rapidly achieves a high-avidity (active) state due to conformational changes in the glycoprotein. These changes in the amount and avidity of CD11/CD18 on the leukocyte surface allows for a strong interaction of the adhesion molecule with its EC counter-receptor (e.g., CD11a/CD18 to ICAM-1 or ICAM-2). The simultaneous rapid up-regulation of CD11b/CD18 and shedding of L-selectin on leukocytes upon activation (Figure 7.3) enables leukocytes to rapidly transition between the rolling and firmly adherent states [120,121,123,140,149].

The pathophysiological importance of selectins, immunoglobulin-like adhesion molecules, and integrins in mediating the proposed adhesive function in intact postcapillary venules is supported

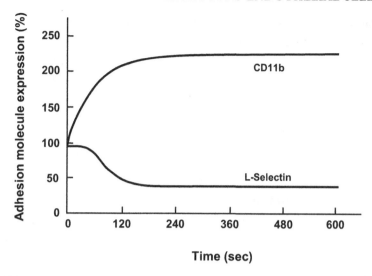

FIGURE 7.3: Time-course of expression of CD11b/CD18 and L-selectin on rabbit neutrophils fol-lowing exposure to platelet activating factor. Within seconds after PAF exposure, the neutrophils shed L-selectin while rapidly increasing the expression of CD11b. Modified from Granger et al. [175]. Re-produced from Physiology of the Gastrointestinal Tract, Granger DN, Grisham MB, Kvietys PR, with permission of Leonard Johnson (Editor).

by a large number of animal studies that employed adhesion molecule-specific blocking monoclonal antibodies (mAbs) and/or adhesion molecule-deficient mice to show an attenuated adhesion re-sponse. In many of these same studies, evidence is also provided to demonstrate the contribution of leukocyte–endothelial cell adhesion to the microvascular dysfunction (e.g., EC barrier dysfunction) and tissue injury that accompanies acute and chronic inflammation [86,127,140,150,151].

7.2   CHEMICAL MEDIATORS
7.2.1   Pro-Adhesive Mediators
A variety of chemical mediators are released from inflamed tissue that can act on receptors expressed on leukocytes and/or ECs to promote or inhibit leukocyte–endothelial cell adhesion. Histamine, leukotrienes, and platelet activating factor (PAF) production/release from degranulated mast cells and/or activated macrophages lying in proximity to the vessel wall are likely signals for the rapid in-duction of leukocyte rolling, with histamine inducing the rapid mobilization of preformed P-selectin the EC surface. The leukotrienes (e.g., LTB_4) and PAF can also activate rolling leukocytes, increase the expression/avidity of CD11/CD18 and initiate the transition to firm adhesion. Cytokines and

chemokines released from perivascular cells and the vessel wall elicit the transcription-dependent expression of endothelial selectin (E- and P-selectin), as well as increasing the expression of ICAM-1 and VCAM-1, which require several hours to achieve peak expression on EC. Activated EC characteristically produce ROS at an accelerated rate, which contributes to transcription-dependent adhesion molecule synthesis/expression by activating key oxidant-sensitive transcription factors (NFkB, AP-1). The changes in EC adhesion molecule expression induced by cytokines and oxidative stress allow for a sustained increase in leukocyte rolling, firm adhesion, and transendothelial migration [120,121].

7.2.2 Anti-Adhesive Mediators

There are several naturally occurring biological agents that appear to serve as endogenous anti-adhesion molecules. These include NO, prostacyclin, and adenosine. Several lines of evidence also implicate NO as an endogenous inhibitor of leukocyte adhesion in venules: (1) NO synthase inhibitors elicit the recruitment of adherent leukocytes, (2) NO donors (nitroprusside, SIN-I) attenuate or prevent the leukocyte adherence induced by different inflammatory stimuli, (3) superoxide, which reacts with NO to render it biologically inactive, promotes leukocyte adherence, and (4) SOD, which scavenges superoxide and limits the inactivation of NO, inhibits leukocyte adhesion. Direct exposure of isolated neutrophils to NO synthase inhibitors does not induce increased expression/avidity of CD11/CD18 nor do they increase the adhesion of leukocytes to biologically inert surfaces (plastic), indicating that these agents promote leukocyte adherence through an action on the endothelium or other cell types. Overall, the available data suggest that any condition that tips the balance between NO production and superoxide generation in favor of the latter will elicit recruitment of adherent leukocytes within postcapillary venules [152–156].

Adenosine and prostacyclin are also very effective in reducing the adherence and emigration of leukocytes in inflamed in postcapillary venules. The anti-adhesive actions of adenosine are mediated through the engagement of A_2, but not A_1, receptors on leukocytes. Although adenosine-mediated inhibition of neutrophilic superoxide production is also mediated through the A_2, receptor, this does not appear to provide the basis for adenosine's anti-adhesive action because SOD is much less effective than adenosine in reducing PAF-induced leukocyte adherence in venules. Adenosine A_2-receptor-mediated effects appear to explain the potent reduction in inflammatory mediator-induced leukocyte adhesion and emigration following treatment with methotrexate, which is commonly used in the treatment of rheumatoid arthritis. Prostacyclin (PGI_2), which is known for its ability to inhibit platelet aggregation, also appears to act as an inhibitor of leukocyte–endothelial cell adhesion. The view that PGI_2 affects leukocyte adherence is supported by the observation that prostaglandin synthesis inhibitors (e.g., indomethacin) promote leukocyte adherence in mesenteric venules, an effect that can be reversed with exogenous PGI. Iloprost, a stable prostacyclin analogue, also exerts a profound inhibitory influence on leukocyte adherence in postcapillary venules exposed to I/R [120,157–163].

7.3   ROLE OF HYDRODYNAMIC FORCES

Shear forces generated by the movement of blood within the microvasculature are generally higher in arterioles than in downstream venules. For example, a 30-μm diameter venules in rat mesentery is likely to exhibit a resting shear rate that is nearly half that of an arteriole of comparable diameter. Since wall shear stress or shear rate represents an anti-adhesion force that opposes the pro-adhesive forces generated by leukocyte and EC adhesion molecules, vessels with a low spontaneous shear rate (low blood flow) would be expected to exhibit more leukocyte adhesion than vessels with high shear rates. For this reason, it has been proposed that leukocytes rarely roll and adhere in arterioles because the higher shear forces exceed the adhesive force in these vessels. Based on this proposal, one might predict that reductions in arteriolar shear rate to levels experienced by venules should promote leukocyte adhesion in arterioles. However, this is not the case because when cat mesenteric arterioles and venules of the same size are exposed to the same range of shear rates (100–1250 sec^{-1}), venules exhibit far more leukocyte rolling and adherence than arterioles. Studies using retrograde perfusion of the microcirculation have also provided some insight into the role of shear rate in the arteriolar–venous differences in leukocyte–endothelial cell adhesion. In the mesentery, retrograde perfusion is associated with a reduced flux of rolling leukocytes in venules and increased leukocyte rolling in arterioles. However, more leukocytes still rolled in venules during normograde perfusion than rolled in arterioles during retrograde flow. These observations indicate that hemodynamic differences between arterioles and venules cannot explain the predilection for leukocyte rolling and adherence in venules and that a more probable explanation for the greater adhesive interactions between leukocytes and venular endothelium is that the counter-receptors (ligands) for leukocyte adhesion molecules are more densely concentrated on venular endothelium [120,140,164–169].

Shear forces can, however, play an important role in modulating the leukocyte–endothelial cell adhesion that occurs during inflammation. The prevailing shear rate exerted on the walls of postcapillary venules determines the level of leukocyte rolling and firm adherence and dictates the contact area between leukocytes and the endothelial cell surface. Even in the absence of an inflammatory stimulus, graded reductions in venular shear rate for brief periods (<2 min) elicit progressive recruitment of both rolling and firmly adherent leukocytes. Similarly, it has been noted that the number of adherent leukocytes recruited into venules by an inflammatory stimulus is inversely proportional to the wall shear rate, suggesting that it is easier for leukocytes to create strong adhesive bonds with ECs at low shear rates and that high shear rates may prevent the creation of such bonds. This effect of venular shear rate on the intensity of the leukocyte–endothelial cell adhesion suggests that the changes in blood flow that are associated with the early and sustained phases of inflammation (discuss above) may exert a significant influence on the intensity of the leukocyte recruitment response during inflammation [165–167].

7.4 CYTOTOXICITY OF ADHERENT LEUKOCYTES

Neutrophils that are firmly adherent to vascular ECs are also activated, which results in the production and release of ROS, proteases, cationic proteins (e.g., defensins) and a variety of other chemicals that can impair the function or inflict injury to microvessels. Activated neutrophils utilize the plasma membrane-associated enzyme NADPH oxidase to produce superoxide, which subsequently reacts with itself (spontaneous dismutation) to generate hydrogen peroxide. The potent oxidizing and chlorinating agent hypochlorous acid (HOCl) is also produced when hydrogen peroxide and extracellular chloride ions react with myeloperoxidase (MPO), a cationic enzyme released from neutrophil granules (Figure 7.4). MPO binds avidly to the negatively charged endothelial glycocalyx and is subsequently internalized, with a resultant rise in intracellular ROS. Internalized MPO modulates vascular signaling and impairs vasodilatory function by decreasing the bioavailability of NO through HOCl-mediated chlorination of L-arginine and direct inactivation of NO [170–176].

Activated neutrophils also secrete a variety of proteases, which have the potential to elicit uncontrolled proteolysis of vascular wall elements (e.g., basement membrane) and of the interstitial matrix. Many of these proteases are secreted in an inactive (latent) form that is dependent on oxidative mechanisms (HOCl) for activation (Figure 7.4). Extracellular fluid is well-endowed with antioxidants and antiproteases, which limit the cytotoxic potential of circulating neutrophils. However, when neutrophils adhere to ECs, a sequestered microenvironment is created (at the leukocyte–endothelial cell interface), which allows neutrophil-derived oxidants and proteases to overwhelm plasma antioxidants and anti-proteases, thereby enabling the neutrophil to exert its full cytotoxic potential at the vessel wall. The accompanying vascular dysfunction can be manifested as a hyperadhesivity of EC to leukocytes and platelets, the formation of microthrombi, increased production of ROS by EC, and diminished endothelial barrier function [174,175].

The major neutrophil-derived proteases include elastase, collagenase, and gelatinase. These enzymes represent potent mechanisms by which neutrophils may degrade the key components of the endothelial cell basement membrane and interstitial matrix. Elastase, like many proteins released from activated neutrophils, is cationic, which facilitates its interaction with the EC surface (glycocalyx). The metalloproteinases (e.g., gelatinase, collagenase) are secreted in a latent, inactive form that requires further processing for activation. HOCl has been shown to oxidatively activate collagenase and gelatinase secreted by human neutrophils. The role for chlorinated oxidants in activating the metalloproteinase is further substantiated by studies that demonstrate that neutrophils from patients with chronic granulomatous disease (neutrophils are unable to generate ROS) are also

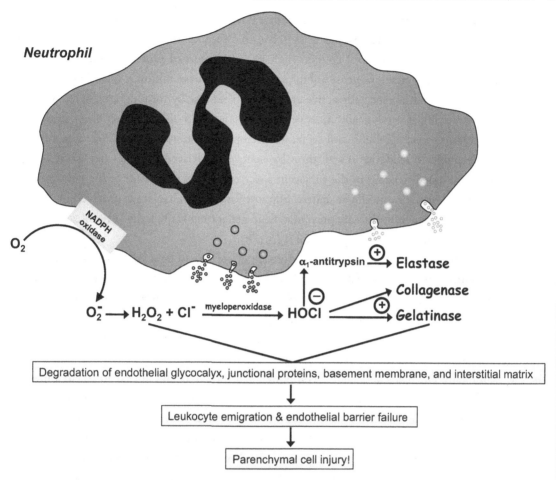

FIGURE 7.4: Neutrophil-mediated mechanisms of microvascular dysfunction/injury. Upon activation, neutrophils generate superoxide (O_2^-) from NADPH oxidase, with a resultant dismutation of O_2^- to form hydrogen peroxide (H_2O_2). The activated neutrophils also degranulate to release MPO and different proteases. MPO reacts with H_2O_2 in the presence of chloride ions to produce HOCl, which facilitates the direct activation of collagenase and gelatinase, while indirectly activating elastase by inhibiting the anti-protease, α_1-antitrypsin. ROS and activated proteases generated from activated, adherent neutrophils can damage the vessel wall, promote leukocyte emigration, and lead to parenchymal cell injury.

unable to activate collagenase. It has also been suggested that activation of gelatinase by chlorinated oxidants is required for neutrophils to degrade the type IV collagen of the endothelial basement membrane and to emigrate from the vasculature. The secretion and subsequent oxidant activation of collagenase would then facilitate the degradation of the interstitial collagens (types I, II, and II), thereby allowing migration of neutrophils within the interstitial matrix [151,172,174,175,177].

The fact that a normal inflammatory response does not result in proteolytic degradation of the interstitium suggests that the host has ways to control this potentially injurious process. Indeed, plasma and interstitial fluid (lymph) contain high concentrations of protease inhibitors. One of best-characterized endogenous protease inhibitors is α_1-proteinase inhibitor (also known as α_1-antitrypsin). This protein is especially active at inhibiting elastase by forming a complex with the protease and rendering it catalytically inactive. Extracellular fluid also contains other proteinase inhibitors such as α_2-macroglobulin and secretory leukoproteinase inhibitor. All of these antiproteases are susceptible to oxidative inactivation by neutrophil-derived oxidants like HOCl. This inactivation is presumed to occur in the subjacent space created by adherence of the neutrophil to cellular membranes or the extracellular matrix. Support for this mechanism is provided by the observation that when the neutrophils are prevented from generating HOCl, the antiproteases remain active and are able to inhibit tissue injury [172–175,178].

The endothelial surface layer (ESL or glycocalyx) appears to be a vulnerable target for the products of neutrophil activation. ROS, MPO, and proteases released by activated adherent neutrophils have the capacity to degrade or depolymerize the membrane-bound glycoproteins and proteoglycans that comprise the glycocalyx. This effectively reduces the thickness of the ESL and creates an opportunity for EC adhesion molecules such as P-selectin to protrude through the ESL and bring the adhesion molecule in close contact with the EC surface. Under normal conditions, the glycosaminoglycan chains and soluble components of the glycocalyx appear to shield adhesion molecules, which prevent the adhesion of circulating cells (leukocyte, platelets) with EC. During inflammation, both EC and adherent neutrophils are activated, resulting in oxidative and enzymatic degradation of the glycocalyx, an opening of the meshwork, and the exposure of EC adhesion molecules, which in turn, amplifies the recruitment of blood cells onto the vessel wall. The leukocyte-mediated ESL degradation also diminishes endothelial barrier function [38–40,179,180].

·   ·   ·   ·

CHAPTER 8

Platelet–Vessel Wall Interactions

Platelets are known to adhere and aggregate at sites of vascular injury in response to endothelial denudation and exposure of subendothelial collagen. The accumulation of platelets at the injury site serves to temporarily plug the damaged vessel and localize subsequent procoagulant events. Recent work has revealed, however, that endothelial denudation is not an absolute requirement for platelet attachment to the walls of blood vessels. Although healthy ECs prevent platelet adhesion by hiding components of the subendothelial matrix (collagen, fibronectin) from platelets, augmenting fibrinolysis, and producing platelet inactivators (NO, PGI_2), inflammation can lead to an altered phenotype of ECs, leukocytes, and platelets that enhances the capacity of these cells to bind to each other. Consequently, platelet–vessel wall interactions are often observed in the inflamed microvasculature. The recruitment of rolling and adherent platelets in blood microvessels appears to be a well-regulated process that involves the expression and/or activation of adhesion molecules on platelets, ECs, and/or leukocytes. Consequently, platelets can bind to the vessel wall either via a direct interaction with ECs or indirectly by attaching to already adherent leukocytes [12,181–187].

A variety of inflammatory stimuli have been shown to elicit the adhesion of platelets in the microcirculation of different vascular beds, including brain, intestine, mesentery, liver, lung, skeletal muscle, and retina. Some of these stimuli induce the rolling and/or firm adhesion of platelets within a few minutes (calcium ionophore A23187, oxidized LDL), while others require hours (endotoxin, TNF-α), or days (*Plasmodium berghei* malaria, hypercholesterolemia [HCh]). In most instances, the platelet adhesion response is confined to the postcapillary venules; however, there are several descriptions of platelet adhesion in arterioles, and fewer descriptions of a response in capillaries [12,137,142,180,184,188–190].

8.1 PLATELET ACTIVATION: MECHANISMS AND CONSEQUENCES

Platelet dysfunction is a feature of acute and chronic inflammatory diseases. This is often manifested as an increased expression of activation-dependent surface antigens on circulating platelets, including P-selectin and CD40 ligand. Soluble CD40L levels in plasma are also elevated, which largely reflects the shedding of this pro-inflammatory signaling molecule from the surface of activated platelets. Changes in platelet function during inflammation are also reflected in the increased tendency for platelets to spontaneously aggregate in vitro [20] and to exhibit an increased sensitivity to endogenous pro-aggregation molecules such as collagen and adenosine diphosphate (ADP). The mechanisms that underlie the platelet activation associated with inflammation remain poorly understood [191–194].

It may be expected that as they course through the microvasculature of inflamed tissue, platelets are likely exposed to a variety of substances that either prime the cells for activation or directly activate them. In regions with tissue damage, platelets may be exposed to collagen, a potent stimulant for activation. The direct contact of platelets with cytokine-activated ECs can also lead to platelet activation. Similarly, a variety of soluble substances released from injured resident cells and/or recruited inflammatory cells may also participate in the activation of platelets within the intestinal vasculature. ADP may accumulate in inflamed tissue either as a result of diminished capillary perfusion or due to inhibition of ectonucleotidase CD39, which is expressed on the surface of EC and circulating immune cells, where it efficiently hydrolyzes extracellular ATP and ADP (both of which stimulate platelet adhesion) to AMP and ultimately adenosine (which inhibits platelet aggregation). Oxidative stress and proinflammatory cytokines (e.g., TNF-α) also down-regulate CD39 on T-lymphocytes. Arachidonic acid and PAF produced in response to phospholipase A_2 activation are also potential mediators of platelet activation in the inflamed tissue. Once platelet activation is initiated, then a variety of substances that are produced (e.g., thromboxane A_2, ADP, serotonin) by platelets, released from granules, or shed (CD40L) from the cell surface can amplify the activation and accumulation of platelets in the microvasculature [187,191,195–197].

An attenuated EC production of endogenous inhibitors of platelet activation, including prostacyclin (PGI_2) and NO, may also contribute to the platelet activation response during inflammation. In vitro studies implicate both NO and superoxide as modulators of homotypic platelet aggregation as well as the adhesion of platelets to monolayers of cultured ECs, with superoxide promoting and NO inhibiting the platelet adhesion responses. Platelets and ECs both produce NO from the constitutive isoform of NO synthase (eNOS) and both cell types as well as leukocytes have the capacity to produce superoxide from NADPH oxidase and other enzymes. NO is known to affect the function of different cells via cyclic GMP (cGMP)-dependent and cGMP-independent (related to target specific nitrosation) pathways. Superoxide scavenging may be an important property of NO that enables it to modulate platelet adhesion during inflammation. NO can react with superoxide at a rate that is three times faster than SOD can convert superoxide to peroxide. This avidity of NO for superoxide (and vice versa) indicates that the balance between NO and superoxide fluxes in microvessels may be an important determinant of how either of these reactive species can modulate platelet adhesion during inflammation [184,198–200].

Both NO and superoxide may also indirectly affect platelet adhesion due to their well established effects on leukocyte–endothelial cell adhesion and VSM tone. Superoxide appears to promote, while NO inhibits, leukocyte–endothelial cell adhesion in postcapillary venules. Since leukocyte-dependent mechanisms contribute heavily to the platelet adhesion observed in some pathophysiological states (as discussed below), an imbalance between NO and superoxide may ultimately affect platelet recruitment by modulating leukocyte adhesion. Similarly, because shear forces generated within microvessels can influence platelet adhesion, the ability of NO to produce vasodilation and for superoxide to cause vasoconstriction could account for some of the altered platelet adhesion responses that are observed when the NO-superoxide balance is altered in inflamed tissue.

Activated platelets produce and release a variety of substances that have the potential to influence the quality and intensity of an inflammatory response. These platelet-derived factors act on both leukocytes and ECs to induce an inflammatory phenotype. Some products of platelet activation contribute to transcellular metabolic reactions in neutrophils, which use arachidonic acid released by platelets to produce increased quantities of inflammatory leukotrienes. The attachment of activated platelets to neutrophils also enables the latter produce larger quantities of superoxide and PAF than either cell is capable of producing alone [201–203].

ECs are also an important target for platelets and their activation products. When CD40L-positive platelets are co-incubated with cultured EC, the ECs become activated, as evidenced by an increased surface expression of ICAM-1 and VCAM-1, an enhanced production of IL-8 (a neutrophil chemoattractant), and increased leukocyte–endothelial cell adhesion. These platelets also release the chemokine RANTES, which binds to glycosaminoglycans on the endothelial cell surface to further promote leukocyte adhesion. Since RANTES has been recently implicated as a mediator of impaired endothelium-dependent vasodilation associated with hypertension, it is possible that this platelet-derived chemokine may contribute to the reduced capacity of arterioles in inflamed tissue to dilate in response to acetylcholine and other endothelium-dependent vasodilators [203–207].

8.2 PLATELET–ENDOTHELIAL ADHESION

In some forms of inflammation, the accumulation of platelets within the microvasculature appears to result from a direct interaction between platelets and vascular ECs. For example, endothelial cell activation appears to play an important role in the platelet adhesion response of cerebral and intestinal venules to HCh. Although circulating platelets also assume an activated phenotype in hypercholesterolemic animals and humans, the adhesion of platelets in the microcirculation occurs only when platelets derived from mice placed on a normal (ND) or high-cholesterol (HCD) diet are monitored in HCD-recipient, but not ND-recipient, mice, suggesting that platelet activation is not sufficient to elicit the adhesion response. Using a similar experimental strategy, endothelial

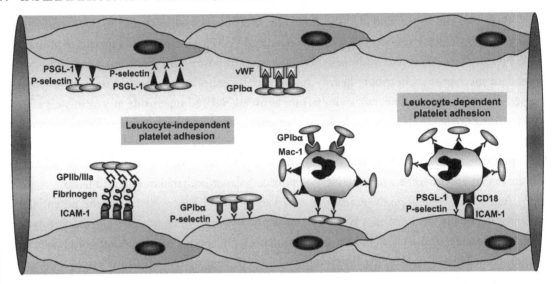

FIGURE 8.1: Leukocyte-dependent and leukocyte-independent mechanisms of platelet adhesion in inflamed postcapillary venules. Platelets may adhere directly to ECs, using glycoprotein receptors, such as GPIbα and GPIIb/IIIa, that bind to ligands that are expressed on the surface of ECs such as vWF, ICAM-1, P-selectin. Platelet P-selectin may also bind to PSGL-1 and unknown ligands expressed on venular endothelium. The adherent platelets may also create a platform onto which leukocytes can roll and establish firm adhesion. Leukocytes can directly roll on endothelial cell P-selectin and then firmly adhere through β2-integrin (CD11/CD18) interactions with endothelial cell ICAM-1. Since adherent leukocytes constitutively express PSGL-1, platelets can bind to these stationary cells via P-selectin–PSGL-1 and GPIbα–CD11b/CD18 interactions. Adapted from Tailor et al. [184]. Reproduced from Platelet-vessel wall interactions in the microcirculation. Microcirculation, Tailor A, Cooper D, Granger DN, with permission of Taylor & Francis.

cell, rather than platelet, activation has also been invoked to explain the platelet adhesion induced by either bacterial endotoxin (lipopolysaccharide [LPS]) or *P. berghei* malaria in WT mice or by hypoxia–reoxygenation in sickle cell transgenic mice [189,190,208–210].

Figure 8.1 illustrates some of the possible adhesion receptor–ligand interactions that may account for the adhesion of platelets to microvascular ECs in different pathophysiological states. Platelet–endothelial cell adhesion elicited in acute models of inflammation (e.g., I/R) has been linked to interactions between ICAM-1 on ECs, fibrinogen, and the platelet adhesion molecule GPIIb/IIIa. In this situation, the oxidative stress experienced by activated EC results in fibrinogen deposition onto constitutively expressed ICAM-1, creating a scaffold on the vessel wall onto which platelets can adhere using GPIIb/IIIa. The inability of platelets derived from patients suffering from Glanzmann's disease, a deficiency in GPIIb/IIIa, to bind to activated EC, as well as the attenuated adhesion response of normal platelets following administration of a fibrinogen or GPIIb/IIIa blocking antibody support a role for ICAM-1:fibrinogen:GPIIb/IIIa mechanism. Direct adhesion of platelets to EC can also result from interactions between platelet associated GP1bα with either P-selectin or von Willebrand factor (vWF) on ECs. PSGL-1 expressed on platelets can also interact with EC P-selectin to mediate adhesion. Similarly, with the recent discovery that EC can express PSGL-1 (discussed above), platelet-associated P-selectin could also bind to PSGL-1 on EC. The shear force in a vessel is also an important determinant of the adhesion molecules that mediate platelet adhesion in inflamed microvessels. Under low shear stress (<600 sec^{-1}), as is seen in venules, the ICAM-1:fibrinogen: GPIIb/IIIa mechanism is an effective mediator of platelet adhesion, while high shear stress, as experienced in arterioles, favors the engagement of platelet GPIba with vWF [181,211–213].

P-selectin has received considerable attention as a determinant of platelet adhesion in inflamed microvessels. This lectin-like adhesion glycoprotein is normally stored in granular structures of both platelets (α-granules) and ECs (Weibel–Palade bodies), from which P-selectin can be rapidly mobilized to the cell surface upon endothelial cell activation. Some vascular beds (e.g., intestine) exhibit significant basal expression of P-selectin, with little or no basal expression on unactivated circulating platelets. Blocking monoclonal antibodies and P-selectin-deficient mice have been used to implicate P-selectin as a mediator of the platelet adhesion (both rolling and firm adhesion) induced by both acute (e.g., A23187, I/R) and chronic (e.g., HCh, *P. berghei* malaria) stimuli. The relative contributions of platelet vs. endothelial cell P-selectin to the platelet adhesion response has been addressed in these and other models of inflammation. This has been achieved using either bone marrow chimeras, produced by the transplantation of bone marrow from P-selectin-deficient donor mice into WT recipient mice (or vice versa) or by monitoring the trafficking of P-selectin-deficient platelets in WT recipients (or vice versa). These experimental strategies have revealed that P-selectin expressed on EC appears to be a major determinant of the platelet adhesion in some inflammation models (e.g., sickle cell disease), while both endothelial cell- and platelet-associated P-selectin contribute to the platelet adhesion in others (e.g., HCh, malaria). In more acute models (e.g., I/R), the rapid adhesion response is entirely dependent on endothelial P-selectin, while the slow, time-dependent platelet adhesion response involves both platelet and endothelial cell P-selectin [184,189,190,204,210,212–215].

8.3 PLATELET–LEUKOCYTE ADHESION

Platelet–vessel wall interactions in inflamed microvessels can also result from the binding of platelets to leukocytes that are already attached to vascular endothelium. Platelet–leukocyte adhesion can be mediated by different ligand–receptor interactions, including P-selectin (platelet)–PSGL-1 (leukocyte) and GPIbα (platelet)–CD11b/CD18 (leukocytes) interactions (Figure 8.1). Evidence supporting a role for leukocytes in platelet adhesion in inflamed venules is provided by studies demonstrating an attenuation of the adhesion response in animals rendered neutropenic. Similarly, an attenuated platelet adhesion response has been demonstrated in mice that are genetically deficient in either CD18 or ICAM-1 and in WT mice receiving blocking antibodies directed against these adhesion glycoproteins. The data from these studies of leukocyte-dependent platelet adhesion are consistent with a model wherein leukocytes require P-selectin to roll on venular endothelium and subsequently establish firm adhesion via a CD18–ICAM-1 interaction. The adherent leukocytes, which express PSGL-1 and other P-selectin ligands, then create a platform onto which platelets can bind using P-selectin. This model would explain why P-selectin expressed on both platelets and ECs is required for platelet adhesion in some experimental models and why interfering with leukocyte adhesion or rendering mice neutropenic can lead to a concomitant reduction in platelet adhesion in the microcirculation [181,184,186,215–217].

Efforts to simultaneously monitor and quantify platelet and leukocyte adhesion have revealed that a significant proportion of platelets adhering in inflamed venules are attached to adherent leukocytes. The percentage of total adherent platelets that are bound to leukocytes vary among models of inflammation. For example, with I/R or HCh as the inflammatory stimulus, approximately 25% of the platelets bind directly to venular endothelium, while the remaining 75% of the adherent platelets are attached to leukocytes that are bound to the vessel wall. In colonic venules of mice with experimental colitis, nearly 100% of the platelets are bound to adherent leukocytes. The platelets that are directly bound to venular endothelium are unaffected by ablation of either ICAM-1 or CD18 function; however, the accumulation of leukocyte-bound platelets is dramatically reduced following the ablation of these leukocyte adhesion receptors. P-selectin blockade, however, effectively attenuates both the leukocyte-dependent and leukocyte-independent components of platelet recruitment. It is noteworthy that approximately 40% to 50% of the leukocytes that adhere in inflamed venules are platelet-bearing and the remaining 50% to 60% are platelet-free, suggesting that a specific subpopulation of the adherent leukocytes may bind platelets. This possibility is supported by evidence that platelets will avidly bind to neutrophils and monocytes, but not lymphocytes. Alternatively, most of the adherent leukocytes may be neutrophils (as suggested by the studies in neutropenic mice) but roughly half of these adherent neutrophils may achieve an activation state that allows for platelet adhesion. This possibility is supported by evidence implicating neutrophil-derived superoxide in the modulation of platelet adhesion in inflamed venules [184,216–218].

8.4   PLATELET–LEUKOCYTE AGGREGATES

A variety of inflammatory diseases are associated with the appearance of platelet-leukocyte aggregates (PLA) in systemic blood. Although PLA formation is not always correlated with disease activity, this heterotypic cell–cell interaction appears to yield platelets that are more intensely activated than their counterparts that participate in homotypic (platelet–platelet) interactions. The increased expression of P-selectin on activated platelets enables these cells to bind to leukocytes, which constitutively express P-selectin glycoprotein ligand-1 (PSGL-1), the major ligand for platelet P-selectin. The P-selectin-dependent platelet–leukocyte complexes that are observed in inflamed venules may be a precursor of the free-flowing platelet–leukocyte aggregates (PLA) that are detected at increased levels in patients with inflammatory diseases. It has been proposed that PLA are initially formed on the endothelial surface of inflamed microvessels, where they are subsequently dislodged by shear forces generated from the movement of blood. Some of the PLA released into venous blood may not appear in systemic circulation due to entrapment of the aggregates in capillaries of the lung and/or liver. Since leukocyte-free P-selectin-positive platelets also appear in systemic blood, then it is also possible that PLA are formed in flowing blood due to engagement of platelet P-selectin with PSGL-1 that is constitutively expressed on leukocytes. Whether leukocyte activation contributes to the formation of PLA remains unclear. However, platelet activators such as thromboxane and platelet activating factor (PAF) generated by the inflamed tissue may predispose platelets to PLA formation. It has been proposed that the PLA may represent an important circulating source of inflammatory mediators that can sustain or amplify an inflammatory response. Leukocytes with attached platelets appear to be primed for adhesion and can achieve a more activated state than their platelet-free counterparts [187,219–224].

CHAPTER 9

Coagulation and Thrombosis

Many inflammatory conditions are associated with a hypercoagulable state and a shift in hemostatic mechanisms in favor of thrombosis. While much attention has been devoted to the formation of potentially lethal thromboemboli in large arteries and veins in inflammatory diseases (e.g., atherosclerosis), there is growing evidence from animal models indicating that inflammation also enhances thrombus formation in the microvasculature. Evidence supporting an influence of inflammation on microvascular thrombosis has been derived from experimental models that compare thrombus formation in control and inflamed microvessels that are subjected to injury induced by mechanical trauma, photoactivation, laser light exposure, electrical stimulation, or topical application of caustic chemicals (e.g., ferric chloride). The underlying mechanism of the platelet–vessel wall interactions and thrombus development associated with these models is dependent on whether the injury response is limited to endothelial cell activation (e.g., photoactivation) or involves endothelial denudation and exposure of the subendothelial collagen (e.g., ferric chloride). Nonetheless, both types of thrombosis models have been used to demonstrate an accelerated thrombosis response in arterioles and/or venules during different acute and chronic models of inflammation. Some of the inflammatory stimuli/models studied to date include bacterial LPS, sepsis induced by cecal ligation and puncture (CLP), DSS-induced colitis, HCh, and hypertension [225–232].

9.1 INTERDEPENDENCE OF COAGULATION AND INFLAMMATION

There is a large and rapidly growing body of evidence suggesting that inflammation and hemostasis are intimately linked processes, wherein each process propagates and intensifies the other, creating the potential for a vicious cycle of thrombogenesis and inflammation (Figure 9.1). The induction of this procoagulant, prothrombotic state likely involves ECs, leukocytes, and platelets, which are activated in response to the inflammatory stimulus. The anticoagulant role of ECs is diminished during inflammation, and this can result from an increased expression of tissue factor (TF, the initiator of coagulation), down-regulation of the anticoagulant protein C pathway, and inactivation of NO by superoxide. The recruitment of rolling and adherent leukocytes on vascular EC can help to create

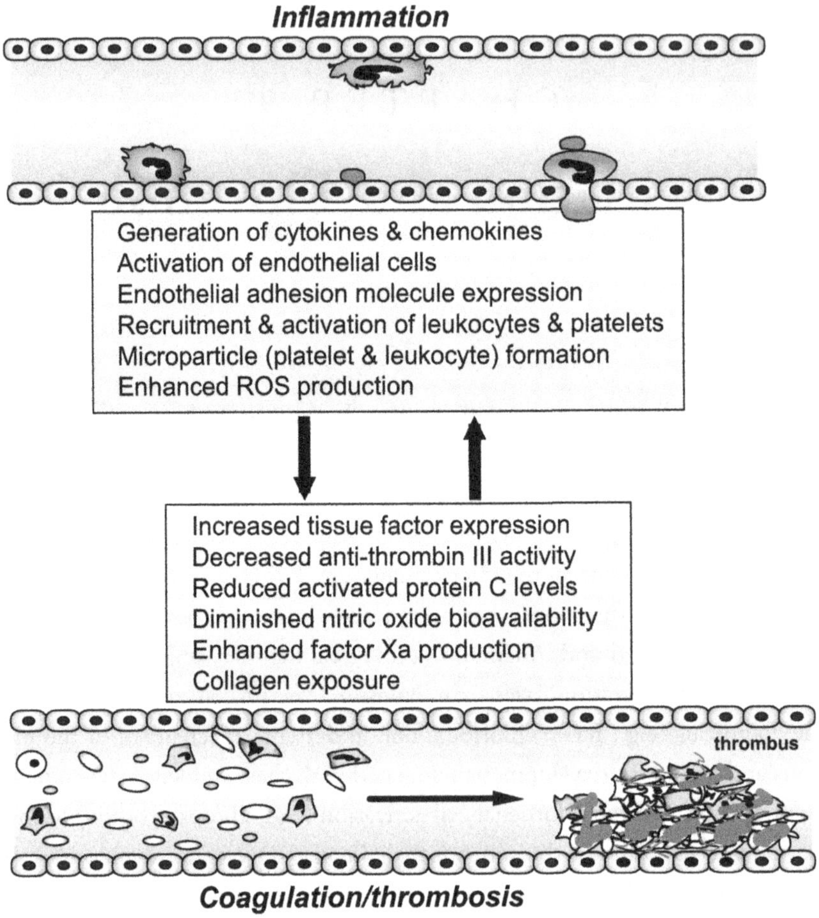

FIGURE 9.1: Vicious cycle of inflammation and coagulation. Inflammation induces a variety of changes in ECs, leukocytes, and platelets, which promote the creation of a procoagulant, prothrombotic surface on the vessel wall (upper). Activation of the coagulation cascade, on the other hand, can elicit changes that promote inflammation (lower). The net result is that inflammation and coagulation/thrombosis are intimately linked processes wherein each process propagates and intensifies the other.

a procoagulant surface for thrombus development. Activated leukocytes also exhibit an increased tissue factor expression and can release proteases that degrade antithrombin as well as cleave and inactivate thrombomodulin on ECs. Tissue factor-rich leukocyte microparticles are also known to contribute to platelet recruitment via P-selectin–PSGL-dependent interactions. The activation and binding of platelets and platelet microparticles to ECs, leukocytes, and to other platelets in the microvasculature of inflamed tissue also promotes a procoagulant state, via an enhanced expression of tissue factor, the generation/activation of coagulation factors (e.g., factor Xa), and enhanced thrombin production [233–238].

In patients with active IBD, for example, there is evidence for accelerated thrombin generation and increased circulating levels of fibrinogen, vWF, thrombin–antithrombin (TAT) complexes, and clotting factors V, VII, and VIII. In addition, antithrombin III, protein C, protein S, plasminogen activating inhibitor (PAI), and tissue plasminogen activator (tPA) levels are often reduced in these individuals. These manifestations of a hypercoagulable state have also been reproduced in animal models of IBD, such as DSS colitis. In both human and experimental IBD, the hemostatic abnormalities are accompanied by functional changes in circulating platelets that exhibit hyperactivity, hyperaggregability, and a propensity to form platelet–leukocyte aggregates [187,193,194,230,239,240–242].

There is also evidence that the coagulation–anticoagulation pathways exert an influence on the inflammatory response. Different components of the coagulation pathways, including thrombin and tissue factor, appear to promote inflammation, while anticoagulants such as activated protein C (APC) and heparin exert anti-inflammatory effects. Platelets, which are recruited to and activated at sites of thrombus formation, also produce and release a myriad of substances that promote inflammation. The influence of hemostasis on inflammation is supported by numerous reports that describe how different components of the coagulation–anticoagulation pathways can regulate inflammation by exerting an influence on ECs, platelets, and/or leukocytes. Thrombin, for example, has been shown to increase the expression (via transcription-independent and transcription-dependent mechanisms) of adhesion molecules on ECs and to promote leukocyte–endothelial cell adhesion. Also, activation of factor XII can result in complement activation. The engagement of TF with its ligand (factor VIIa) activates the protease-activated receptors PAR1–4, which elicits the production of pro-inflammatory cytokines (TNF-α, IL-6), increases the expression of EC adhesion molecules, and promotes leukocyte rolling in venules. Mice that lack the cytoplasmic domain of TF exhibit an attenuated recruitment of rolling, adherent, and transmigrating leukocytes in postcapillary venules after LPS challenge. Furthermore, a small molecule inhibitor of the TF-VIIa complex (BCX-3607) has been shown to attenuate LPS-induced production of IL-6 and IL-8 in vitro (by ECs) and IL-6 in vivo. Similarly, APC has been shown to inhibit the production of adhesion molecules (VCAM-1, ICAM-1) and cytokines in ECs, as well as agonist-induced leukocyte activation and LPS-induced production of TNF-α and other cytokines by cultured monocytes/macrophages. Finally, mice with single-allele targeted disruption of the protein C gene (heterozygous protein C deficient (PC$^{+/-}$) mice) have higher levels of circulating cytokines, including a fourfold increase in TNF-α, after endotoxin challenge, which is consistent with an anti-inflammatory action of APC [233,243–246].

9.2 INFLAMMATION-INDUCED MICROVASCULAR THROMBOSIS: SITE-SPECIFIC RESPONSES

While both arterioles and venules exhibit the capacity for thrombus formation, some inflammatory stimuli appear to preferentially enhance thrombogenesis in one segment of the microvasculature. For example, venules are far more responsive than arterioles to the thrombosis-enhancing effects of endotoxin (LPS) or sepsis induced by CLP. DSS colitis, angiotensin II-induced hypertension, and HCh, on the other hand, appear to preferentially enhance thrombus formation in arterioles. The basis for this predisposition of arterioles and venules to certain inflammatory stimuli/conditions remains unclear. However, there are a number of characteristic differences in the function and behavior of the two vascular segments that could underlie the differential thrombogenic responses. Thrombi formed in the venous system are characteristically rich in fibrin and trapped red blood cells and poor in platelets, while arterial thrombi are rich in aggregated platelets. Arterioles exhibit a higher shear rate than venules, which would favor the participation of some coagulation factors (e.g., vWF) in the thrombogenic response in arterioles. The higher shear rate in arterioles may also render these vessels vulnerable to inflammation-induced reductions in NO bioavailability. Studies of the concentration profile of platelets within arterioles and venules have revealed that the density of platelets near the vessel wall is much higher in arterioles than in venules, and this difference in platelet distribution cannot be attributed to the occurrence of leukocyte margination in venules [225,228–231,247–249].

Venules differ from arterioles in other ways that could explain the differential thrombogenic responses to inflammatory stimuli. It is likely that the density, distribution and/or production of pro- and anti-coagulation factors (e.g., TF) differs between the vessel types and that venules exhibit less dilution of locally generated procoagulants (e.g., thrombin) by the slower moving blood. The preferential trafficking of leukocytes on venular endothelium during inflammation may render venules more vulnerable to certain inflammatory stimuli. For example, it has been reported that neutrophils may promote thrombosis via the release of neutrophil extracellular traps (NETs) [44]. Bacterial endotoxin as well as activated platelets can induce neutrophils to make NETs in the microvasculature. NETs are abundant in thrombi associated with deep vein thrombosis, and it has been shown that platelets under flow in vitro bind avidly to NETs and are able to promote thrombosis. It has been suggested that the backbone of NETs, which is made of chromatin, provides a structure upon which platelets can adhere, become activated, and aggregate, thereby contributing to thrombus initiation and/or stability. Whether NET formation accounts for the enhanced venular thrombosis induced by LPS remains unclear; however, reports describing no effect of either neutropenia or immuno-blockade of adhesion molecules that mediate leukocyte–endothelial cell adhesion in response to LPS on the venular thrombosis response would argue against a role for neutrophil-derived NETs in this model of inflammation-enhanced thrombosis [249–253].

9.3   CHEMICAL MEDIATORS OF INFLAMMATION-ENHANCED THROMBOSIS

While different cell populations (platelets, leukocytes, ECs) and microparticles derived from some of these cells (platelets, leukocytes) have been implicated as factors linking inflammation to thrombosis, a role for chemical mediators that are generated in response to activation of these and other inflammatory cells also appears likely. The involvement of circulating blood cells and/or soluble mediators is supported by reports describing enhanced thrombus formation in tissues distant from the inflammatory site. For example, colonic inflammation in mice results in accelerated thrombosis in arterioles of the cremaster muscle. This may merely reflect the passage of blood cells (e.g., monocytes) and/or microparticles that are activated to produce TF as they course through the gut circulation and eventually transit through extra-intestinal vascular beds, such as skeletal muscle. Alternatively, proteins released (cytokines, chemokines) or shed (e.g., soluble forms of CD40L or the endothelial protein C receptor, EPCR) from EC and/or inflammatory cells within inflamed tissue could mediate the distant (as well as the local) thrombotic response to inflammation [187,241].

Pro-inflammatory cytokines may mediate the enhanced microvascular thrombosis that is associated with inflammation. Several cytokines, including IL-1β, TNF-α, and IL-6, appear to be powerful inducers of coagulation. IL-1β, TNF-α, and IL-6 are known to enhance the expression of tissue factor on ECs and monocytes, down-regulate thrombomodulin, reduce the density of endothelial protein C receptors, and inhibit fibrinolysis on ECs. These cytokine is also known to elicit the shedding of EPCR, thereby producing circulating soluble EPCR that can inhibit protein C activation. TNF-α is also known to increase plasma levels of vWF and to deplete tissue factor pathway inhibitor. Incubation of endothelial cell monolayers with purified recombinant TNF-α elicits a time- and dose-dependent increase in tissue factor procoagulant activity. The cytokine also acts directly on ECs to release both tissue- (tPA) and urokinase-type (uPA) plasminogen activators, while also increasing plasminogen activator inhibitor (PAI-1), with inhibition of fibrinolysis and inadequate removal of fibrin as the net result. Indirectly, TNF-α can lead to impaired anticoagulation by causing leukocytes to release elastase, which could cleave and inactivate thrombomodulin on vascular ECs and antithrombin III. Finally, the ectonucleotidase CD39, which is expressed on ECs, platelets and leukocytes, represents another potential target for the prothrombotic actions of cytokines. TNF-α and oxidative stress are known to down-regulate CD39 expression on leukocytes, which would blunt the efficient hydrolysis of extracellular ATP and ADP, and limit the production of adenosine, which is antithrombotic. Transgenic mice that overexpress CD39 are protected against thrombosis. Consequently, an attenuated expression and/or activity of CD39, which has been described in hypertension, IBD, and HCh, may account for the enhanced thrombus formation in acute and chronic inflammatory states [233–236,254–256–259].

Another member of the TNF superfamily of inflammatory molecules, the CD40/CD40L signaling pathway, may also provide a link between inflammation and coagulation/thrombosis. This pathway has been implicated in the microvascular recruitment of platelets in animal models of IBD and HCh. Upon activation, platelets shed large quantities of CD40 ligand (CD40L), which can promote thrombosis by engaging with its receptor on EC to elicit the increased biosynthesis and expression of EC adhesion molecules and induce tissue factor-dependent procoagulant activity. Soluble CD40L (sCD40L), of which >95% of the circulating level is derived from platelets, also acts as a ligand for the platelet glycoprotein GPIIb/IIIa, and the engagement of CD40L with GPIIb/IIa helps to stabilize the thrombus stabilization and activate more platelets. This contention is supported by reports describing delayed arteriolar thrombosis following vessel injury with $FeCl_3$ in CD40L-deficient mice and the restoration of arteriolar thrombosis when the $CD40L^{-/-}$ mice received recombinant sCD40L. The possibility that CD40L contributes to the hypercoagulable, prothrombotic state in chronic inflammatory diseases is also supported by evidence for increased CD40L expression on circulating platelets and increased soluble CD40L (sCD40L) in plasma of patients with atherosclerosis, rheumatoid arthritis, and IBD. Indeed, sCD40L levels are used as a prognostic marker of thrombotic risk in cardiovascular disease [192,195,196,260,261].

9.4 REACTIVE OXYGEN AND NITROGEN SPECIES

Enhanced ROS production and diminished NO bioavailability may also contribute to the enhanced thrombosis that accompanies inflammation. NO inhibits platelet function and prevents thrombosis, while ROS (particularly superoxide) promotes platelet aggregation and thrombosis. A role for ROS in thrombus development has been demonstrated in cerebral arterioles subjected to photoactivation (which generates ROS). In this model, platelet thrombus formation was inhibited by dimethyl sulfoxide (DMSO), a hydroxyl radical scavenger, and by SOD. Thrombus formation induced by iontophoretic application of ADP on mesenteric venules is greatly attenuated by SOD treatment and, to a lesser extent, by catalase administration. The nonselective NOS inhibitor L-NAME, enhanced ADP-induced thrombus growth while L-arginine administration had no effect. NOS inhibition also enhances thrombus development in arterioles and venules injured by photoactivation, while NO donors and NO-independent guanylate cyclase activators (e.g., YC-1) have been reported to dose-dependently inhibit thrombus development in the microvasculature. The enhanced arteriolar thrombosis that accompanies HCh is completely reversed by topical delivery of L-arginine (the substrate for NO production by eNOS). While NO appears to afford protection against thrombus development in both arterioles and venules, it appears that endogenous NO production is more important in inhibiting thrombus development in venules than in arterioles [262–267].

CHAPTER 10

Endothelial Barrier Dysfunction

ECs normally serve as a barrier to the movement of fluid and proteins from the intravascular compartment to the interstitium. When this barrier function is diminished, either as a consequence of endothelial cell damage or contraction of adjacent ECs, plasma proteins gain greater access to the interstitial compartment, resulting in an elevated oncotic pressure and the withdrawal of fluid from the intravascular compartment, an increased interstitial fluid volume, and interstitial edema. While capillaries are the major source of fluid that is filtered into the interstitial spaces, postcapillary venules represent the major site of vascular protein leakage (extravagation). The role of venules in protein extravasation is particularly evident in inflamed tissue because the accumulated inflammatory mediators and immune cells can act on venular ECs to diminish barrier function [12,15,19,22,151,268–270].

10.1 SITE OF INFLAMMATION-INDUCED BARRIER FAILURE

There are several characteristic features of postcapillary venules that enables this segment of the microvasculature to regulate vascular permeability and the rate of egress of plasma proteins into the interstitium. Ultrastructural analyses of the pathways for transvascular exchange have revealed that both the size and frequency of interendothelial junctions and endothelial fenestrae are higher in postcapillary venules than in either arterioles or capillaries. These pathways are normally large enough to allow a low basal amount of plasma protein leakage that is driven by both diffusive and convective (coupled to fluid filtration) mechanisms. Venular endothelium also appears to possess a higher density of cell surface receptors for inflammatory mediators than their counterparts in arterioles and capillaries. Engagement of certain inflammatory mediators (e.g., histamine, PAF) with their receptors on venular ECs elicits subtle changes in the fine structure of the endothelial monolayer, such as a widening of the endothelial paracellular junctions, which results from the dissociation of junctional proteins and/or cytoskeletal contraction, and a consequent increase in the rate of protein extravasation. Furthermore, since postcapillary venules are the preferred site for leukocyte and platelet adhesion, venular ECs are more frequently and directly exposed to products of leukocyte (e.g., elastase, ROS) and platelets (e.g., RANTES, CD40L), which can diminish barrier function, compared to ECs in arterioles and capillaries [15,22,268–270].

The permeation of albumin across an intact endothelial barrier can occur through transcellular as well as paracellular pathways (Figure 10.1). Protein transcytosis via plasmalemmal vesicles containing caveoli-1 has been demonstrated in ECs of different vascular beds, and it has been ascribed a significant role in the transport of macromolecules across continuous type capillaries. The vesicles shuttle plasma proteins between blood and the interstitium, producing a diffusive flux that favors net transport into the interstitium. While the quantitative importance of vesicular transport to total protein transport remains controversial, recent attention on vesicles has been directed toward their ability to interconnect with each other to produce grape cluster-like structures (vesiculovacuolar organelles, VVOs) that function as open channels that connect (and transport proteins between) the blood and interstitial compartments. Since the increased vascular permeability induced by some inflammatory mediators, including histamine and VEGF, is associated with an increased number of VVOs, it has been proposed that this pathway may contribute to the endothelial barrier dysfunction that accompanies inflammation [22,271,272].

The exchange of plasma proteins through interendothelial cell junctions (the paracellular pathway) is generally considered the primary target of the chemical mediators that elicit the endothelial barrier failure associated with inflammation. The restrictive properties (barrier function) of

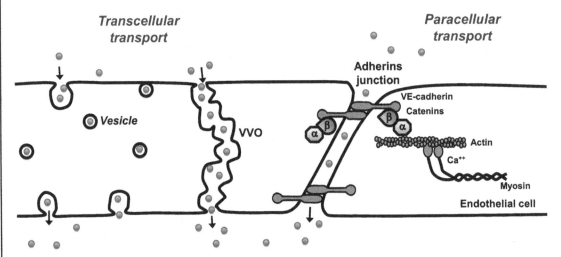

FIGURE 10.1: Routes of solute exchange across the inflamed endothelial barrier. Albumin and other plasma proteins can gain access to the interstitial compartment from plasma via either a paracellular (junctional) or intracellular (vesicular) transport pathway. The integrity of the adherens junction is regulated by junctional proteins such as VE-cadherin, which is linked to actin components of the cytoskeleton via the catenins. The paracellular pathway includes plasmalemmal vesicles that shuttle plasma proteins between the blood and interstitial interface of the endothelial cell, as well as patent channels formed by the fusion of vesicle clusters called *vesiculovacuolar organelles* (VVO).

the endothelial cell junctions (adherens junctions) in most vascular beds result from the homophilic binding of VE cadherin expressed on the surface of two adjacent EC (Figure 10.1). The intracellular domains of VE cadherin molecules are anchored to the EC cytoskeleton via actin-binding proteins called *catenins*. The more highly restrictive cerebral microvessels (blood brain barrier) possess specialized (tight) junctions wherein the cell–cell contact area is maintained by other junctional proteins including claudins, junctional adhesion molecules, occludins, and zonular occludins. Inflammatory mediators that increase vascular permeability in microvessels with adherens junctions exert this effect by disrupting junctional complex assembly via the phosphorylation, internalization, and/or degradation of junctional molecules. The engagement of these mediators with their EC receptors elicits an increased cytosolic calcium and/or the activation of different signaling cascades, leading to cytoskeletal contraction, a loss of junctional protein stability and binding interactions, junctional complex disassembly, and ultimately endothelial barrier failure. Adherens junctions also support the transendothelial migration of leukocytes during inflammation. Consequently, once the inflammatory response resolves, i.e., mediator release subsides and leukocyte transmigration ceases, resealing of the junctional pathways is initiated to restore normal endothelial barrier properties [268–270,273].

A variety of different experimental approaches have been used to assess endothelial barrier dysfunction in vivo and in vitro models of inflammation. Changes in the restrictive properties of monolayers of cultured ECs have been monitored using measurements of transendothelial resistance (cerebral ECs) or the permeability coefficient for albumin or other solutes. Isolated perfused venules have also been used to monitor changes in albumin permeability coefficients and consequently barrier function. In vivo approaches include measurements of the osmotic reflection coefficient for plasma proteins using lymph-to-plasma protein concentration ratios, single vessel estimates of hydraulic conductivity, electron microscopic evaluation of horseradish peroxidase leakage, and quantifying the extravasation of Evans blue dye or fluorescently labeled macromolecules [49]. Data generated with these methods are generally interpreted as reflecting changes in vascular permeability; however, caution should be given to the possibility that, under some circumstances, increased solute extravasation may reflect an increased diffusive exchange due to capillary recruitment and/or increased exchange due to convection, without a corresponding change in the restrictive properties of the endothelial barrier [274–278].

10.2   ROLE OF CIRCULATING BLOOD CELLS

10.2.1 Leukocytes

Leukocytes produce and secrete a variety of factors that are capable of increasing vascular permeability (Table 10.1). The adhesion and transendothelial migration of leukocytes in inflamed venules have been linked to endothelial barrier dysfunction in both acute and chronic models of inflammation. There are several lines of evidence implicating neutrophils in inflammation-induced

TABLE 10.1: Activation products released by leukocytes and platelets that may impair endothelial barrier function.

LEUKOCYTES	PLATELETS
Reactive oxygen species	*Reactive oxygen species*
Superoxide	Superoxide
Hydrogen peroxide	Hydrogen peroxide
Proteases	*Growth factors*
Cathepsin-G	PDGF
Elastase	Transforming growth factor-β
Collagenase	VEGF
Cytokines/chemokines	*Cytokines/chemokines*
IL-1, IL-2, IL-6, IL-8, IL-12	IL-1, IL-7, IL-8
IFN-α, IFN-β	RANTES (regulated upon activation,
TNF-α, TNF-β	normal T cell expressed and secreted)
Transforming growth factor-β	TNF-β
Monocyte chemotactic factor-1	CD40 ligand
Oxidases	*Lipid mediators*
Myeloperoxidase	Thromboxane A_2
	12-HETE
Lipid mediators	
Leukotrienes B4, C4	*Procoagulants*
Platelet activating factor	Thrombin
	ADT and ATP
Miscellaneous	Platelet factor-4
Cationic proteins	Polyphosphates
Histamine	
VEGF	

Modified from Rodrigues and Granger [316].

permeability responses. For example, animals that are rendered neutropenic or that receive blocking antibodies that prevent neutrophil–endothelial cell adhesion exhibit a blunted vascular permeability response in different models of inflammation. Similarly, mice that are genetically deficient in key endothelial cell (e.g., ICAM-1, P-selectin) or leukocyte (CD11/CD18) adhesion molecules also show an improved endothelial barrier function during inflammation. Indirect evidence supporting a role for neutrophils is provided by reports describing a significant positive correlation between the magnitude of the inflammation-associated increase in vascular permeability and the number of adherent and/or emigrated neutrophils in/around postcapillary venules [204,275,279–284].

Whether transendothelial cell migration per se is a requirement for the leukocyte-dependent permeability response remains unclear. However, there is evidence suggesting that proteases released from activated adherent leukocytes can degrade (and promote the internalization) of EC junctional proteins. Furthermore, leukocyte adhesion-induced signaling events in ECs (e.g., increased intracellular calcium) have been implicated in the barrier dysfunction induced by inflammatory mediators. For example, engagement of leukocyte integrins with ICAM-1 or VCAM-1 on EC can result in junctional protein phosphorylation and internalization as well as stimulate the production of ROS through activation of endothelial cell NADPH oxidase. Since transendothelial leukocyte migration is not always associated with endothelial barrier dysfunction, whether an inflammatory mediator elicits an increased vascular permeability is likely determined by the level of activation (release of proteases and ROS) of the leukocyte as it traverses the junction and if adhesion-dependent signaling mechanisms cause EC activation, junctional protein disassembly, and cytoskeletal contraction [24,177,270,285,286].

While neutrophils have received more attention as mediators of endothelial barrier dysfunction during inflammation, there is also growing evidence for the involvement of T-lymphocytes, which are recruited into inflamed microvessels hour to days after the appearance of neutrophils. The pathophysiological relevance of the T-cell recruitment is evidenced by the observation that immunodeficient SCID mice (which lack T-cells) exhibit a significantly blunted vascular permeability response to acute or chronic inflammation and that the permeability response can be restored to WT levels when SCID mice are reconstituted with T-cells from WT mice [11]. Antibody-induced depletion of either CD4+ or CD8+ T-cells has also been used to implicate these T-cell populations in inflammation-induced vascular permeability responses. In some inflammation models, the permeability response appears to be dependent on T-cells, yet there is no evidence for increased trafficking of the immune cells through the affected microvessels. This suggests that T-cell activation products (e.g., cytokines) may mediate the barrier dysfunction via a mechanism that does not require T-cell–endothelial cell adhesion. The T-cell-derived mediators may directly act on the EC to cause barrier failure or may promote the recruitment and/or activation of other cells (e.g., neutrophils, mast cells) that exert a more direct influence on EC [287–289].

10.2.2 Platelets

Like leukocytes, platelets produce a large number of chemical agents (e.g., VEGF, ROS, thrombin) that have the potential to impair endothelial barrier function (Table 10.1). Nonetheless, exposing monolayers of cultured ECs to platelets or platelet-conditioned media typically results in enhanced barrier function, i.e., a reduction in vascular permeability. Experimental evidence suggests that a sphingolipid (sphingosine-1-phosphate, S1P), which is released by activated platelets, accounts for this barrier protective effect of platelets and it acts by reorganizing the actin cytoskeleton of EC. While there are few reports that directly address the role of platelets in inflammation-induced barrier dysfunction, some of these studies suggest that the presence of platelets either does not alter or improves endothelial barrier function during acute inflammation. For example, one study of the postischemic coronary vasculature indicates that platelets are not requisite for the barrier dysfunction, while another reports describes an improved barrier function following the addition of platelets. A protective effect of platelets has also been reported for the lung. A limitation of all these studies is that the organs used to assess the permeability responses to the inflammatory insult were perfused with artificial solutions, to which platelets are added. Since the attachment of platelets to neutrophils is known to enhance the capacity of the leukocytes to produce ROS, it is conceivable that the vascular permeability enhancing potential of platelets is not realized in the absence of other blood cells [290–294].

10.3 ROLE OF PERIVASCULAR CELLS

Mast cells residing in the perivascular space have also been implicated in the endothelial barrier dysfunction that accompanies inflammation. Mast cell degranulation, a common feature of the inflammatory response, is characterized by the production and release of superoxide, amines (histamine, serotonin), leukotrienes, proteases, and cytokines (e.g., TNF-α, IL-1), all of which can diminish endothelial barrier function. A role for mast cell degranulation in inflammation-induced endothelial barrier failure is supported by reports describing an attenuating influence of mast cell stabilizing drugs or a genetic deficiency of mast cells on the vascular permeability response during acute and chronic inflammation. The microvasculature in mast cell-rich tissues, such as the intestine and lung, typically exhibit a dependence on mast cells in the increased vascular permeability response to inflammation. Since the myriad of factors released during mast cell degranulation can exert an influence on the recruitment of other inflammatory cells into inflamed microvessels, it is often difficult to distinguish between a direct effect of mast cell products on endothelial barrier function vs. the actions of secondary cells (e.g., leukocytes) that are recruited or activated by those products. However, some mediators that are relatively unique to mast cells (e.g., histamine, mast cell-specific serine proteases) have been shown to contribute to inflammation-induced permeability responses in a variety of animal models. Histamine typically exerts its barrier-altering action on EC by engaging the Gq-coupled H1 receptor, which elevates intracellular calcium, and triggers actin–myosin contraction. Mast cell-specific tryptases and chymases promote vascular permeability via indirect (enhancing bradykinin production) and possibly direct mechanisms. Even tissues with a relatively low population of mast cells, such as the brain, show evidence of mast cell dependent barrier failure. For example, the impaired blood brain barrier function that occurs in brain following focal ischemia/reperfusion is significantly blunted in WT rats treated with a mast cell stabilizer (cromoglycate) and in mast cell deficient rats [17,57,58,295–298].

Macrophages that reside in the perivascular compartment have also been implicated in inflammation-induced endothelial barrier failure. Activated macrophages also produce and release a variety of cytokines, chemokines as well as ROS and NO, which can impair endothelial barrier function. Macrophage depletion can be achieved in vivo by intravascular injection of agents that are selectively cytotoxic to macrophages, such as clodronate liposomes or 2-chloroadenosine. Using these reagents, a dependency of endothelial barrier function on macrophages has been described in tissues with a large resident population of macrophages (e.g., lung, gut) and in tissues with smaller resident populations (e.g., nerves) [299–303].

10.4   REACTIVE OXYGEN AND NITROGEN SPECIES

ECs, leukocytes, platelets, macrophages, and mast cells are activated and produce ROS at an accelerated rate during inflammation. The elevated ROS levels have been proposed as potential mediators of inflammation-induced endothelial barrier failure. ROS can exert this effect in a variety of ways. Direct mechanisms include phosphorylation of catenins with the subsequent dissociation of VE-cadherins, eliciting actin–myosin cytoskeletal contraction, and degradation of the endocapillary layer (glycocalyx). Indirect actions of ROS include enhanced leukocyte adhesion and transendothelial migration via oxidant sensitive, transcription-dependent up-regulation of EC adhesion molecules, and activation of latent proteases such as MMPs. Lipid peroxidation-mediated cell membrane damage induced by the elevated ROS fluxes is an unlikely contributor to the inflammation-mediated changes in barrier function [13,39,40,268,270,273].

A role for ROS in the vascular permeability responses to inflammation is supported by studies that have focused on inhibiting the production of ROS by targeting enzymatic sources such as xanthine oxidase or NADPH oxidase, as well as studies that examine the influence of ROS scavengers (SOD) on the permeability response. These approaches have been used to implicate xanthine oxidase-derived ROS in the enhanced microvascular protein and water permeability induced by I/R. Since xanthine oxidase inhibition or ROS scavenging has also been shown to largely prevent the adherence and transendothelial migration of leukocytes in I/R and other inflammatory states, it remains unclear from these observations whether the ROS directly alter impair barrier function or do so indirectly by limiting the adhesive interactions between leukocytes and ECs [13,175,304,305].

Another source of ROS that has received considerable attention in inflamed vessels is NADPH oxidase. Nonspecific NADPH oxidase inhibitors as well as mice that are genetically deficient in critical protein subunits (e.g., p47phox, gp91phox) of the enzyme complex have been used to implicate this enzyme as a major source of ROS that mediates endothelial barrier failure.

Chimeras produced by the transplantation of bone marrow from p47^{phox} deficient into WT recipients (or vice versa) have revealed that bone marrow-derived cells, rather than ECs, mediate the NADPH oxidase-dependent barrier alterations elicited in the pulmonary microvasculature. Focal brain I/R also results in impaired endothelial barrier function that is dependent on NADPH oxidase, as evidenced by experiments showing reduced blood brain barrier dysfunction after focal ischemia stroke and reperfusion in NADPH oxidase (gp91^{phox}) deficient mice and in WT mice treated with apocynin [306–309].

An important pathophysiological consequence of increased superoxide production in inflamed ECs is inactivation of NO. As detailed above, physiologic levels of NO appear to play an important role in preventing leukocyte–endothelial cell, platelet–endothelial cell, and platelet–leukocyte adhesion in the normal noninflamed microvasculature, and it serves to stabilize mast cells in the perivascular space. Upon inactivation of NO by superoxide, the cell–cell interactions are elicited, mast cell degranulate, and there is a reduction in endothelial barrier function. NO donors have been shown to blunt all of these responses to increased ROS production. Similar protection against ROS-induced endothelial barrier failure has been noted in mice that genetically overexpress endothelial NO synthase (eNOS). In contrast, nonselective inhibition of NOS in otherwise healthy tissue results in oxidative stress and increased vascular permeability. While these studies suggest that NO normally protects against endothelial barrier failure, there is also evidence indicating that NO directly diminishes barrier function. The incongruent permeability responses to NO can be explained by a study that compared the water permeability responses of mesenteric venules perfused with or without blood-borne constituents to NOS inhibition. It was noted that in the absence of blood-borne constituents, permeability was reduced by approximately 50%, while a >75% increase in permeability was detected in vessels perfused by blood during exposure to the NOS inhibitor. Hence, these findings suggest that the major pathophysiological consequence of a reduction in NO bioavailability in inflamed microvessels is the loss of a critical anti-adhesion molecule that serves to limit the barrier compromising effects of leukocyte adhesion and transendothelial migration [310–314].

· · · ·

Epilogue

The preceding discussion underscores the complexity of the inflammatory response and the diversity in the responses of the microvasculature to this condition. The large and continually expanding body of knowledge on the microcirculation and its contribution to inflammation has revealed the participation of all segments of the microvasculature, i.e., arterioles, capillaries, and venules. The best characterized phenotypic changes that are exhibited by the microvasculature during inflammation include impaired vasomotor function, reduced capillary perfusion, adhesion of leukocytes and platelets, activation of the coagulation cascade and enhanced thrombosis, increased vascular permeability, and an increase in the rate of proliferation of blood and lymphatic vessels. Collectively, these responses appear to be geared toward enhancing the delivery of inflammatory cells to the injured/infected tissue, isolating the region from healthy tissue and the systemic circulation, and setting the stage for tissue repair and regeneration.

It appears that virtually every cell that either resides within or courses through the inflamed region is activated, which allows the cells that have a stake in the outcome of the inflammatory response to make a meaningful contribution. Although these cells are highly specialized, they respond to an inflammatory response in much the same way, i.e., by producing excess amounts of ROS or by releasing cytokines and/or proteases. This redundancy likely serves to amplify the response and achieve an intensity that ensures rapid eradication of an infectious agent. It remains unclear whether any single cell type plays a dominant role in coordinating the microvascular response to inflammation. If such an orchestrator exists, it is likely the endothelial cell. These cells serve as the gatekeeper for the migration of inflammatory cells into the affected tissue, modulate the tone of underlying VSM and hence blood flow, provide an interface for the activation/deposition of coagulation factors and binding of platelets during thrombus formation, and play a critical role in the formation of new blood vessels. While little attention has been devoted to if/how the responses of arterioles, capillaries, and venules are coordinated during inflammation, ECs also appear to be well suited to assume such a function.

Although we have conveniently addressed the microvascular responses to inflammation as separate, independent processes, there is compelling evidence that these events are closely linked and share many chemical mediators and signaling pathways. A growing number of mediators have been identified that have the capacity to elicit most or all of the characteristic vascular responses to inflammation, which supports the view that redundant reinforcing signals are received along the entire length of the microvasculature. Thrombin, for example, which has received most attention for its critical role in the coagulation/thrombosis response, is also able to modulate vasomotor function, leukocyte recruitment, vascular permeability, and angiogenesis. VEGF and some cytokines (e.g., TNF-α) are additional examples of single mediators that can elicit most (if not all) of the characteristic microvascular responses to inflammation. The production, release and/or actions of these pleiotropic mediators also appear to be interdependent. For example, thrombin promotes VEGF production and induces an increased expression of VEGF receptors, while TNF-α can elicit thrombin production. Perhaps the single most important chemical alteration in inflamed tissue that accounts for the subsequent generation of a whole host of inflammatory mediators is the imbalance between ROS and NO species. The fact that virtually every cell involved in the inflammatory response undergoes an oxidative burst underscores the importance of ROS in this process. ROS, like soluble inflammatory mediators, exert pleiotropic actions that serve to further amplify the inflammatory response.

The recent discovery of the multitude of mediators and signaling mechanisms that contribute to the inflammatory response has raised hopes for the development of effective new drugs for the treatment of inflammatory diseases. Drugs that target a specific microvascular response to inflammation, such as leukocyte–endothelial cell adhesion or angiogenesis, have shown promise in both the preclinical and clinical studies of inflammatory disease. Future research efforts in this area are certain to identify new avenues for therapeutic intervention and should reveal whether agents that target a single chemical mediator or multiple mediators with overlapping actions show the most promise for clinical success.

References

[1]   Libby P, Ridker PM, Hansson GK. Inflammation in atherosclerosis: from pathophysiology to practice. *J Am Coll Cardiol.* 2009; 54: 2129–38.

[2]   Zeyda M, Stulnig TM. Obesity, Inflammation, and insulin resistance—a mini-review. *Gerontology.* 2009; 55(4): 379–86.

[3]   Medzhitov R. Origin and physiological roles of inflammation. *Nature.* 2008; 454: 428–435.

[4]   Carden DL, Granger DN. Pathophysiology of ischaemia-reperfusion injury. *J Pathol.* 2000; 190: 255–66.

[5]   Querfurth HW, LaFerla FM. Alzheimer's disease. *N Engl J Med.* 2010; 362(4): 329–44.

[6]   Rocha e Silva M. A brief survey of the history of inflammation. 1978. *Agents Actions.* 1994; 43: 86–90.

[7]   Cotran RS. Inflammation: Historical perspectives. In: *Inflammation: Basic principles and clinical correlates.* 3rd Edition (Gallin JI & Snyderman R, eds.), Chapt. 1, pp. 5–10, Lippincott Williams & Wilkins, Philadelphia, 1999.

[8]   Ley K. History of inflammation research. In: *Physiology of Inflammation* (Ley K, ed.), Chapt. 1, pp. 1–10, Oxford Univ Press, 2001.

[9]   Harlan J, Liu DY. In vivo models of leukocyte adherence to endothelium. In: *Adhesion: Its role in inflammatory disease.* Chapt. 6, pp. 117–76, W.H. Freeman & Co., New York, 1992.

[10]   Hurley JV. Inflammation. In: *Edema* (Staub NC & Taylor AE, eds.), Chapt. 19, pp. 463–88, Raven Press, New York, 1984.

[11]   Ley K. The microcirculation in inflammation. In: *Handbook of Physiology: Microcirculation* (Tuma RF, Duran WN, & Ley K, eds.), Chapt. 9, pp. 387–448, Academic Press, San Diego, 2008.

[12]   Granger DN, Rodrigues SF, Yildirim A, Senchenkova EY. Microvascular responses to cardiovascular risk factors. *Microcirculation.* 2010; 17: 192–205.

[13]   Granger DN. Ischemia-reperfusion: mechanisms of microvascular dysfunction and the influence of risk factors for cardiovascular disease. *Microcirculation.* 1999; 6: 167–78.

[14]   Wiedman MP. Architecture. In: *Handbook of Physiology. Section 2: The Cardiovascular System, Vol. IV, Microcirculation, Part 1* (Renkin EM & Michel CC, eds.), Chapt. 2, 11–40, American Physiological Society, Bethesda, Md., 1984.

[15]   Arfors KE, Rutili G, Svenjo E. Microvascular transport of macromolecules in normal and inflammatory conditions. *Acta Physiol Scand Suppl.* 1979; 463: 93–103.

[16]   Hirschi KK, D'Amore PA. Pericytes in the microvasculature. *Cardiovasc Res.* 1996; 32: 687–98.

[17]   Kubes P, Granger DN. Leukocyte-endothelial cell interactions evoked by mast cells. *Cardiovasc Res.* 1996; 32: 699–708.

[18]   Grant L. The sticking and emigration of white blood cells in inflammation. In: *The inflammatory process* (Zweifach B, Grant L, & McCluskey L, eds.), Vol. 2, p. 205, Academic Press, Orlando, 1972.

[19]   Landis EM, Pappenheimer JR. Exchange of substances through capillary walls. In: *Handbook of Physiology*, Section 2; Circulation (Hamilton WF & Dow P, eds.), Chapt. 29, pp. 961–1034, American Physiological Society, Washington, DC, 1963.

[20]   Harlan J. Leukocyte-endothelial interactions. *Blood.* 1985; 65: 513–25.

[21]   Majno G. Chronic inflammation. Links with angiogenesis and wound healing. *Am J Pathol.* 1998; pp. 1035–39.

[22]   Palade GE, Simionescu M, Simionescu N. Structural aspects of the permeability of the microvascular endothelium. *Acta Physiol Scand Suppl.* 1979: 463: 11–32.

[23]   Aird WC. The role of the endothelium in severe sepsis and multiple organ dysfunction syndrome. *Blood.* 2003; 101(10): 3765–77.

[24]   Kvietys PR, Granger DN. Endothelial cell monolayers as a tool for studying microvascular pathophysiology. *Am J Physiol.* 1997; 273: G1189–99.

[25]   Ellis CG, Jagger J, Sharpe M. The microcirculation as a functional system. *Crit Care.* 2005; 9 Suppl 4: S3–8.

[26]   Johnson PC. Overview of the microcirculation. In: *Handbook of Physiology: Microcirculation*, (Tuma RF, Duran WN, & Ley K, eds.), pp. xi–xxiv, Academic Press, San Diego, 2008.

[27]   Kessel RG, Karden RH. *Tissues and organs: A text-atlas of scanning electron microscopy.* W.H. Freeman & Co, San Francisco, 1979.

[28]   Granger DN. Physiology and pathophysiology of the microcirculation. *Dialogues Cardiovasc Med.* 1988; 3: 123–40.

[29]   Kuo L, Davis ML, Chilian WM. Endothelial modulation of arteriolar tone. *News Physiol Sci.* 1992; 7: 5–9.

[30]   Chillan WM. Coronary circulation in health and disease. Summary of an NHLBI Workshop. *Circulation.* 1997; 95: 522–28.

[31]   Granger HJ, Shepherd AP. Intrinsic microvascular control of tissue oxygen delivery. *Microvasc Res.* 1973; 5: 49–72.

[32]   Pries AR, Kuebler WM. Normal endothelium. *Handb Exp Pharmacol.* 2006; 176: 1–40.

[33]   Iigo Y, Suematsu M, Higashida T, Oheda J, Matsumoto K, Wakabayashi Y, Ishimura Y, Miyasaka M, Takashi T. Constitutive expression of ICAM-J in rat microvascular systems analyzed by laser confocal microscopy. *Am J Physiol.* 1997; 273: 138–47.

[34]   Silverstein, Roy L. (1991). The Vascular Endothelium. In *Inflammation: Basic Principles and Clinical Correlates*, 3rd edition by John I. Gallin and Ralph Snyderman. Lippincott Williams & Wilkins (pp. 207–25), Philadelphia.

[35]   Gerritsen ME. Functional heterogeneity of vascular endothelial cells. *Biochem Pharmacol.* 1987; 36: 2701–11.

[36]   Gerritsen ME, Bloor CM. Endothelial cell gene expression in response to injury. *FASEB J.* 1993; 7: 523–32.

[37]   Langer HF, Chavakis T. Leukocyte-endothelial interactions in inflammation. *J Cell Mol Med.* 2009; 13: 1211–20.

[38]   Reitsma S, Slaaf DW, Vink H, van Zandvoort MA, oude Egbrink MG. The endothelial glycocalyx: composition, functions, and visualization. *Pflugers Arch.* 2007; 454: 345–59.

[39]   Pries AR, Secomb TW, Gaehtgens P. The endothelial surface layer. *Pflugers Arch.* 2000; 440: 653–66.

[40]   Flessner MF. Endothelial glycocalyx and the peritoneal barrier. *Perit Dial Int.* 2008; 28: 6–12.

[41]   Grisham MB, Granger DN, Lefer DJ. Modulation of leukocyte-endothelial interactions by reactive metabolites of oxygen and nitrogen: relevance to ischemic heart disease. *Free Radic Biol Med.* 1998; 25: 404–33.

[42]   Grisham MB, Jourd'Heuil D, Wink DA. Nitric oxide. I. Physiological chemistry of nitric oxide and its metabolites: implications in inflammation. *Am J Physiol.* 1999; 276: G315–21.

[43]   Stokes KY, Cooper D, Tailor A, Granger DN. Hypercholesterolemia promotes inflammation and microvascular dysfunction: role of nitric oxide and superoxide. *Free Radic Biol Med.* 2002; 33: 1026–36.

[44]   Ridnour LA, Thomas DD, Mancardi D, Espey MG, Miranda KM, Paolocci N, Feelisch M, Fukuto J, Wink DA. The chemistry of nitrosative stress induced by nitric oxide and reactive nitrogen oxide species. Putting perspective on stressful biological situations. *Biol Chem.* 2004; 385: 1–10.

[45]   Shusta EV. Blood-brain barrier. In: *Endothelial cells in health and disease* (Aird WC, ed.), Taylor and Francis, Boca Raton, Chapt. 2, pp. 33–64, 2005.

[46]   Granger DN, Stokes KY. Differential regulation of leukocyte-endothelial cell adhesion.

In: *Endothelial cells in health and disease* (Aird WC, ed.), Taylor and Francis, Boca Raton, Chapt. 13, pp. 229–44, 2005.

[47] Laughlin MH. Endothelium-mediated control of coronary vascular tone after chronic exercise training. *Med Sci Sports Exerc.* 1995; 27: 1135–44.

[48] Davis MJ, Hill MA, Kuo L. Local regulation of microvascular perfusion. In: *Handbook of Physiology: Microcirculation* (Tuma RF, Duran WN, & Ley K, eds.), Chapt. 6 pp. 161–284, Academic Press, San Diego, 2008.

[49] Tran CH, Welsh DG. The differential hypothesis: a provocative rationalization of the conducted vasomotor response. *Microcirculation.* 2010; 17: 226–36.

[50] Csiszar A, Lehoux S, Ungvari Z. Hemodynamic forces, vascular oxidative stress, and regulation of BMP-2/4 expression. *Antioxid Redox Signal.* 2009; 11: 1683–97.

[51] Zhang C. The role of inflammatory cytokines in endothelial dysfunction. *Basic Res Cardiol.* 2008; 103(5): 398–406.

[52] Kutcher ME, Herman IM. The pericyte: cellular regulator of microvascular blood flow. *Microvasc Res.* 2009; 77: 235–46.

[53] Armulik A, Abramsson A, Betsholtz C. Endothelial/pericyte interactions. *Circ Res.* 2005; 97: 512–23.

[54] Bergers G, Song S. The role of pericytes in blood-vessel formation and maintenance. *Neuro Oncol.* 2005; 7: 452–64.

[55] Kubes P, Granger DN. Leukocyte-endothelial cell interactions evoked by mast cells. *Cardiovasc Res.* 1996; 32: 699–708.

[56] Crivellato E, Travan L, Ribatti D. Mast cells and basophils: a potential link in promoting angiogenesis during allergic inflammation. *Int Arch Allergy Immunol.* 2010; 151: 89–97.

[57] Ryan JJ, Fernando JF. Mast cell modulation of the immune response. *Curr Allergy Asthma Rep.* 2009; 9: 353–59.

[58] Kalesnikoff J, Galli SJ. New developments in mast cell biology. *Nat Immunol.* 2008; 9: 1215–23.

[59] Szekanecz Z, Koch AE. Macrophages and their products in rheumatoid arthritis. *Curr Opin Rheumatol.* 2007; 19: 289–95.

[60] Buckley CD, Pilling D, Lord JM, Akbar AN, Scheel-Toellner D, Salmon M. Fibroblasts regulate the switch from acute resolving to chronic persistent inflammation. *Trends Immunol.* 2001; 22: 199–204.

[61] Sirén V, Salmenperä P, Kankuri E, Bizik J, Sorsa T, Tervahartiala T, Vaheri A. Cell–cell contact activation of fibroblasts increases the expression of matrix metalloproteinases. *Ann Med.* 2006; 38: 212–20.

[62] Kvietys PR, Granger DN. Physiology and pathophysiology of the colonic circulation. *Clin Gastroenterol.* 1986; 15: 967–83.

[63]   Hultén L, Lindhagen J, Lundgren O, Fasth S, Ahrén C. Regional intestinal blood flow in ulcerative colitis and Crohn's disease. *Gastroenterology*. 1977; 72: 388–96.

[64]   Holzer P. Neurogenic vasodilatation and plasma leakage in the skin. *Gen Pharmacol*. 1998; 30: 5–11.

[65]   Hatoum OA, Miura H, Binion DG. The vascular contribution in the pathogenesis of inflammatory bowel disease. *Am J Physiol Heart Circ Physiol*. 2003; 285: H1791–96.

[66]   Vanhoutte PM. Endothelial control of vasomotor function: From health to coronary disease. *Circ J*. 2003; 67: 572–75.

[67]   Feletou M, Tang EH, Vanhoutte PM. Nitric oxide the gatekeeper of endothelial vasomotor control. *Front Biosci*. 2008; 13: 4198–217.

[68]   Mori M, Stokes KY, Vowinkel T, Watanabe N, Elrod JW, Harris NR, Lefer DJ, Hibi T, Granger DN. Colonic blood flow responses in experimental colitis: time course and underlying mechanisms. *Am J Physiol Gastrointest Liver Physiol*. 2005; 289: G1024–29.

[69]   Deniz M, Cetinel S, and Kurtel H. Blood flow alterations in TNBS-induced colitis: role of endothelin receptors. *Inflamm Res*. 2004; 53: 329–36.

[70]   Garrelds IM, Heiligers JP, Van Meeteren ME, Duncker DJ, Saxena PR, Meijssen MA, and Zijlstra FJ. Intestinal blood flow in murine colitis induced with dextran sulfate sodium. *Dig Dis Sci*. 2002; 47: 2231–36.

[71]   Satoyoshi K, Akita Y, Nozu F, Yoshikawa N, and Mitamura K. Hemodynamics in the colonic mucosa of rats with dextran sulfate-induced colitis in the early phase. *J Gastroenterol*. 1996; 31: 512–17.

[72]   Harris NR, Carter PR, Lee S, Watts MN, Zhang S, Grisham MB. Association between blood flow and inflammatory state in a T-cell transfer model of inflammatory bowel disease in mice. *Inflamm Bowel Dis*. 2010; 16: 776–82.

[73]   Hatoum OA, Binion DG, Otterson MF, Gutterman DD. Acquired microvascular dysfunction in inflammatory bowel disease: Loss of nitric oxide-mediated vasodilation. *Gastroenterology*. 2003; 125: 58–69.

[74]   Feletou M. Calcium-activated potassium channels and endothelial dysfunction: therapeutic options? *Br J Pharmacol*. 2009; 156: 545–62.

[75]   Gonzalez MA, Selwyn AP. Endothelial function, inflammation, and prognosis in cardiovascular disease. *Am J Med*. 2003; 115 Suppl 8A: 99S–106S.

[76]   Davignon J, Ganz P. Role of endothelial dysfunction in atherosclerosis. *Circulation*. 2004; 109 (23 Suppl 1): III27–32.

[77]   Huang AL, Vita JA. Effects of systemic inflammation on endothelium-dependent vasodilation. *Trends Cardiovasc Med*. 2006; 16: 15–20.

[78]   Vila E, Salaices M. Cytokines and vascular reactivity in resistance arteries. *Am J Physiol Heart Circ Physiol*. 2005; 288: H1016–21.

[79] Banda MA, Lefer DJ, Granger DN. Postischemic endothelium-dependent vascular reactivity is preserved in adhesion molecule-deficient mice. *Am J Physiol*. 1997; 273: H2721–25.

[80] Lehr HA, Bittinger F, Kirkpatrick CJ. Microcirculatory dysfunction in sepsis: a pathogenetic basis for therapy? *J Pathol*. 2000; 190: 373–86.

[81] Adams DH, Nash GB. Disturbance of leucocyte circulation and adhesion to the endothelium as factors in circulatory pathology. *Br J Anaesth*. 1996; 77: 17–31.

[82] Skalak R, Skalak TC. Flow behavior of leukocytes in small tubes. In: *Physiology & pathophysiology of leukocyte adhesion* (DN Granger & GW Schmid-Schonbein, eds.), Oxford University Press, New York, 1995, Chapt. 4, pp. 97–115.

[83] Hansell P, Borgstrom P, Arfors KE. Pressure-related capillary leukostasis following ischemia-reperfusion or hemorrhagic shock. *Am J Physiol*. 1993; 265: H381–88.

[84] Harris AG, Skalak TC. Leukocyte cytoskeletal structure determines capillary plugging and network resistance in skeletal muscle. *Am J Physiol*. 1993; 265: H1670–75.

[85] Horie Y, Wolf R, Miyasaka M, Anderson DC, Granger DN. Leukocyte adhesion and hepatic microvascular responses to intestinal ischemia/reperfusion in rats. *Gastroenterology*. 1996; 111(3): 666–73.

[86] Horie Y, Wolf R, Anderson DC, Granger DN. Hepatic leukostasis and hypoxic stress in adhesion molecule-deficient mice after gut ischemia/reperfusion. *J Clin Invest*. 1997; 99: 781–88.

[87] Horie Y, Wolf R, Russell J, Shanley TP, Granger DN. Role of Kupffer cells in gut ischemia/reperfusion-induced hepatic microvascular dysfunction in mice. *Hepatology*. 1997; 26: 1499–05.

[88] Horie Y, Wolf R, Chervenak RP, Jennings SR, Granger DN. T-lymphocytes contribute to hepatic leukostasis and hypoxic stress induced by gut ischemia-reperfusion. *Microcirculation*. 1999; 6: 267–80.

[89] Jerome SN, Kong L, Korthuis RJ. Microvascular dysfunction in postischemic skeletal muscle. *J Invest Surg*. 1994; 7: 3–16.

[90] Engler RL, Dahlgren MD, Morris DD, Peterson MA, Schmid-Schonbein GW. Role of leukocytes in the response to acute myocardial ischemia and reflow in dogs. *Am J Physiol*. 1986; 251: H314–22.

[91] Jerome SN, Akimitsu T, Korthuis RJ. Leukocyte adhesion, edema, and development of postischemic capillary no-reflow. *Am J Physiol*. 1994; 267: H1329–36.

[92] Koutroubakis IE, Tsiolakidou G, Karmiris K, Kouroumalis EA. Role of angiogenesis in inflammatory bowel disease. *Inflamm Bowel Dis*. 2006; 12: 515–23.

[93] Lingen MW. Role of leukocytes and endothelial cells in the development of angiogenesis in inflammation and wound healing. *Arch Pathol Lab Med*. 2001; 125: 67–71.

[94] Arenberg DA, Strieter RM. Angiogenesis. In *Inflammation: Basic Principles and Clinical*

Correlates (pp. 851–864), 3rd edition edited by John I. Gallin and Ralph Snyderman. Lippincott Williams & Wilkins, Philadelphia, 1991.

[95]   Nagy JA, Benjamin L, Zeng H, Dvorak AM, Dvorak HF. Vascular permeability, vascular hyperpermeability and angiogenesis. *Angiogenesis*. 2008; 11: 109–19.

[96]   David Dong ZM, Aplin AC, Nicosia RF. Regulation of angiogenesis by macrophages, dendritic cells, and circulating myelomonocytic cells. *Curr Pharm Des*. 2009; pp. 365–79.

[97]   Naldini A, Carraro F. Role of inflammatory mediators in angiogenesis. *Curr Drug Targets Inflamm Allergy*. 2005; 4: 3–8.

[98]   Majno G. Chronic inflammation: links with angiogenesis and wound healing. *Am J Pathol*. 1998; 153: 1035–39.

[99]   Flister MJ, Wilber A, Hall KL, Iwata C, Miyazono K, Nisato RE, Pepper MS, Zawieja DC, Ran S. Inflammation induces lymphangiogenesis through up-regulation of VEGFR-3 mediated by NF-kappaB and Prox1. *Blood*. 2010; 115: 418–29.

[100] Dvorak HF. Angiogenesis: update 2005. *J Thromb Haemost*. 2005; 3: 1835–42.

[101] Shibuya M. Brain angiogenesis in developmental and pathological processes: therapeutic aspects of vascular endothelial growth factor. *FEBS J*. 2009; 276: 4636–43.

[102] Scaldaferri F, Vetrano S, Sans M, Arena V, Straface G, Stigliano E, Repici A, Sturm A, Malesci A, Panes J, Yla-Herttuala S, Fiocchi C, Danese S. VEGF-A links angiogenesis and inflammation in inflammatory bowel disease pathogenesis. *Gastroenterology*. 2009; 136: 585–95.

[103] Chidlow JH Jr, Langston W, Greer JJ, Ostanin D, Abdelbaqi M, Houghton J, Senthilkumar A, Shukla D, Mazar AP, Grisham MB, Kevil CG. Differential angiogenic regulation of experimental colitis. *Am J Pathol*. 2006; 169: 2014–30.

[104] Chidlow JH Jr, Shukla D, Grisham MB, Kevil CG. Pathogenic angiogenesis in IBD and experimental colitis: new ideas and therapeutic avenues. *Am J Physiol Gastrointest Liver Physiol*. 2007; 293: G5–G18.

[105] Nieves BJ, D'Amore PA, Bryan BA. The function of vascular endothelial growth factor. *Biofactors*. 2009; 35: 332–37.

[106] Ng YS, Krilleke D, Shima DT. VEGF function in vascular pathogenesis. *Exp Cell Res*. 2006; 312: 527–37.

[107] Hicklin DJ, Ellis LM. Role of the vascular endothelial growth factor pathway in tumor growth and angiogenesis. *J Clin Oncol*. 2005; 23: 1011–27.

[108] Keeley EC, Mehrad B, Strieter RM. Chemokines as mediators of neovascularization. *Arterioscler Thromb Vasc Biol*. 2008; 28: 1928–36.

[109] Balestrieri ML, Balestrieri A, Mancini FP, Napoli C. Understanding the immunoangiostatic CXC chemokine network. *Cardiovasc Res*. 2008; 78: 250–56.

[110]   Strieter RM, Belperio JA, Phillips RJ, Keane MP. CXC chemokines in angiogenesis of cancer. *Semin Cancer Biol.* 2004; 14: 195–200.

[111]   Ushio-Fukai M, Urao N. Novel role of NADPH oxidase in angiogenesis and stem/progenitor cell function. *Antioxid Redox Signal.* 2009; 11: 2517–33.

[112]   Ushio-Fukai M, Alexander RW. Reactive oxygen species as mediators of angiogenesis signaling: role of NAD(P)H oxidase. *Mol Cell Biochem.* 2004; 264: 85–97.

[113]   Sessa WC. Molecular control of blood flow and angiogenesis: role of nitric oxide. *J Thromb Haemost.* 2009; 7 Suppl 1: 35–37.

[114]   Chidlow JH Jr, Greer JJ, Anthoni C, Bernatchez P, Fernandez-Hernando C, Bruce M, Abdelbaqi M, Shukla D, Granger DN, Sessa WC, Kevil CG. Endothelial caveolin-1 regulates pathologic angiogenesis in a mouse model of colitis. *Gastroenterology.* 2009; 136: 575–84.

[115]   Tammela T, Alitalo K. Lymphangiogenesis: Molecular mechanisms and future promise. *Cell.* 2010; 140: 460–76.

[116]   Saharinen P, Tammela T, Karkkainen MJ, Alitalo K. Lymphatic vasculature: development, molecular regulation and role in tumor metastasis and inflammation. *Trends Immunol.* 2004; 25: 387–95.

[117]   Takahashi M, Yoshimoto T, Kubo H. Molecular mechanisms of lymphangiogenesis. *Int J Hematol.* 2004; 80: 29–34.

[118]   Lohela M, Bry M, Tammela T, Alitalo K. VEGFs and receptors involved in angiogenesis versus lymphangiogenesis. *Curr Opin Cell Biol.* 2009; 21: 154–65.

[119]   Alexander JS, Ganta VC, Jordan PA, Witte MH. Gastrointestinal lymphatics in health and disease. *Pathophysiology* 2009 (Dec 16.) [Epub ahead of print]

[120]   Granger DN, Kubes P. The microcirculation and inflammation: modulation of leukocyte-endothelial cell adhesion. *J Leukoc Bioi.* 1994; 55: 662–75.

[121]   Panés J, Granger DN. Leukocyte-endothelial cell interactions: molecular mechanisms and implications in gastrointestinal disease. *Gastroenterology.* 1998; 114: 1066–90.

[122]   Petri B, Phillipson M, Kubes P. The physiology of leukocyte recruitment: an in vivo perspective. *J Immunol.* 2008; 180: 6439–46.

[123]   Gavins F, Yilmaz G, Granger DN. The evolving paradigm for blood cell-endothelial cell interactions in the cerebral microcirculation. *Microcirculation.* 2007; 14: 667–81.

[124]   Chen L, Lin SX, Amin S, Overbergh L, Maggiolino G, Chan LS. VCAM-1 blockade delays disease onset, reduces disease severity and inflammatory cells in an atopic dermatitis model. *Immunol Cell Biol.* 2010; 88: 334–42.

[125]   Kum WW, Lee S, Grassl GA, Bidshahri R, Hsu K, Ziltener HJ, Finlay BB. Lack of functional P-selectin ligand exacerbates Salmonella serovar typhimurium infection. *J Immunol.* 2009; 182: 6550–61.

[126] Woollard KJ, Suhartoyo A, Harris EE, Eisenhardt SU, Jackson SP, Peter K, Dart AM, Hickey MJ, Chin-Dusting JP. Pathophysiological levels of soluble P-selectin mediate adhesion of leukocytes to the endothelium through Mac-1 activation. *Circ Res.* 2008; 103: 1128–38.

[127] Kakkar AK, Lefer DJ. Leukocyte and endothelial adhesion molecule studies in knockout mice. *Curr Opin Pharmacol.* 2004; 4: 154–58.

[128] Bienvenu K, Granger DN, Perry MA. Flow dependence of leukocyte-endothelial cell adhesion in postcapillary venules. In: *Physiology and Pathophysiology of Leukocyte Adhesion.* (Granger DN & Schmid-Schonbein G, eds.), Oxford University Press, New York, 1995, pp. 278–93.

[129] Steinhoff G, Behrend M, Schrader B, Duijvestijn AM, Wonigeit K. Expression patterns of leukocyte adhesion ligand molecules on human liver endothelia. Lack of ELAM-1 and CD 62 inducibility on sinusoidal endothelia and distinct distribution of VCAM-1, ICAM-1, ICAM-2, and LFA-3. *Am J Pathol.* 1933; 142: 481–88.

[130] Watanabe K, Suematsu M, Iida M, Takashi K, Iisuka Y, Suzuki H, Suzuki M, Tsuchiya M, Tsurufuji S. Effect of rat CINC/gro, a member of the interleukin-8 family, on leukocytes in microcirculation of the rat mesentery. *Exp Mol Pathol.* 1992; 56: 60–69.

[131] Keelan ETM, License ST, Peters AM, Binns RM, Haskard DO. Characterization of E-selectin expression in vivo with use of a radiolabeled monoclonal antibody. *Am J Physiol.* 1994; 266: H279–90.

[132] Eppihimer MJ, Wolitzky B, Anderson DC, Labow MA, Granger DN. Heterogeneity of expression of E- and P-selectins in vivo. *Circ Res.* 1996; 79: 560–69.

[133] Eppihimer MJ, DN Granger. Endothelial cell adhesion molecule expression in acutely inflamed tissues. In: *Yearbook of Intensive Care and Emergency Medicine* (Vincent JL, ed.). Springer-Verlag, Berlin: 1997, pp. 52–62.

[134] Bauer P, Welbourne T, Shigematsu T, Russell J, Granger DN. Endothelial expression of selectins during endotoxin preconditioning. *Am J Physiol Reg Integr Comp Physiol.* 2000; 279: R2015–21.

[135] Hickey MJ, Kanwar S, McCafferty D-M, Granger DN, Eppihimer MJ, Kubes P. Varying Roles of E-selectin and P-selectin in different microvascular beds in response to antigen. *J Immunol.* 1999; 162: 1137–43.

[136] Liu L, Kubes P. Molecular mechanisms of leukocyte recruitment: organ-specific mechanisms of action. *Thromb Haemost.* 2003; 89: 213–20.

[137] Carvalho-Tavares J, Hickey MJ, Hutchison J, Michaud J, Sutcliffe IT, Kubes P. A role for platelets and endothelial selectins in tumor necrosis factor-alpha-induced leukocyte recruitment in the brain microvasculature. *Circ Res.* 2000; 87: 1141–48.

[138] Eppihimer MJ, Russell J, Anderson DC, Epstein CJ, Laroux S, Granger DN. Modulation of P-selectin expression in the postischemic intestinal microvasculature. *Am J Physiol.* 1997; 273: G1326–32.

[139] McEver RP. Selectins: lectins that initiate cell adhesion under flow. *Curr Opin Cell Biol.* 2002; 14: 581–86.

[140] Ley K, Laudanna C, Cybulsky MI, Nourshargh S. Getting to the site of inflammation: the leukocyte adhesion cascade updated. *Nat Rev Immunol.* 2007; 7: 678–89.

[141] Laszik Z, Jansen PJ, Cummings RD, Tedder TF, McEver RP, Moore KL. P-selectin glycoprotein ligand-1 is broadly expressed in cells of myeloid, lymphoid, and dendritic lineage and in some nonhematopoietic cells. *Blood.* 1996; 88: 3010–21.

[142] Vowinkel T, Wood KC, Stokes KY, Russell J, Tailor A, Anthoni C, Senninger N, Krieglstein CF, Granger DN. Mechanisms of platelet and leukocyte recruitment in experimental colitis. *Am J Physiol Gastrointest Liver Physiol.* 2007; 293: G1054–60.

[143] Rivera-Nieves J, Burcin TL, Olson TS, Morris MA, McDuffie M, Cominelli F, Ley K. Critical role of endothelial P-selectin glycoprotein ligand 1 in chronic murine ileitis. *J Exp Med.* 2006; 203: 907–17.

[144] Panés J, Perry MA, Anderson DC, Manning A, Leone B, Cepinskas G, Rosenbloom CL, Miyasaka M, Kvietys PR, Granger DN. Regional differences in constitutive and induced ICAM-1 expression in vivo. *Am J Physiol.* 1995; 269: H1955–64.

[145] Henninger DD, Panés J, Eppihimer M, Russell J, Gerritsen M, Anderson DC, Granger DN. Cytokine-induced VCAM-1 and ICAM-1 expression in different organs of the mouse. *J Immunol.* 1997; 158: 1825–32.

[146] Komatsu S, Berg RD, Russell JM, Nimura Y, Granger DN. Enteric microflora contribute to constitutive ICAM-1 expression on vascular endothelial cells. *Am J Physiol Gastrointest Liver Physiol.* 2000; 279: G186–91.

[147] Komatsu S, Flores S, Gerritsen ME, Anderson DC, Granger DN. Differential Upregulation of circulating soluble and endothelial cell intercellular adhesion molecule-1 in mice. *Am J Pathol.* 1997; 151: 205–14.

[148] Constans J, Conri C. Circulating markers of endothelial function in cardiovascular disease. *Clin Chim Acta.* 2006; 368: 33–47.

[149] Rahman A, Fazal F. Hug tightly and say goodbye: role of endothelial ICAM-1 in leukocyte transmigration. *Antioxid Redox Signal.* 2009; 11: 823–39.

[150] Granger DN, Stokes KY, Shigematsu T, Cerwinka WH, Tailor A, Krieglstein CF. Splanchnic ischaemia-reperfusion injury: mechanistic insights provided by mutant mice. *Acta Physiol Scand.* 2001; 173: 83–91.

[151] DiStasi MR, Ley K. Opening the flood-gates: how neutrophil-endothelial interactions regulate permeability. *Trends Immunol.* 2009; 30: 547–56.

[152] Hickey MJ, Kubes P. Role of nitric oxide in regulation of leucocyte-endothelial cell interactions. *Exp Physiol.* 1997; 82: 339–48.

[153] Suzuki M, Inauen W, Kvietys PR, Grisham MB, Meininger C, Schelling ME, Granger HJ, Granger DN. Superoxide mediates reperfusion-induced leukocyte-endothelial cell interactions. *Am J Physiol.* 1989; 257: H1740–45.

[154] Kubes P, Suzuki M, Granger DN. Nitric oxide: an endogenous modulator of leukocyte adhesion. *Proc Natl Acad Sci USA.* 1991; 88: 4651–55.

[155] Kubes P, Kurose I, Granger DN. NO donors prevent integrin-induced leukocyte adhesion but not P-selectin-dependent rolling in postischemic venules. *Am J Physiol.* 1994; 267: H931–37.

[156] Kurose I, Wolf R, Grisham MB, Aw TY, Specian RD, Granger DN. Microvascular responses to inhibition of nitric oxide production. Role of active oxidants. *Circ Res.* 1995; 76: 30–39.

[157] Cronstein BN, Levin RI, Belanoff J, Weissman G, Hirschhorn R. Adenosine: an endogenous inhibitor of neutrophil-mediated injury to endothelial cells. *J Clin Invest.* 1986; 78: 760–70.

[158] Cronstein BN, Levin RI, Philips M, Hinschhorn R, Abramson SB, Weissman G. Neutrophil adherence to endothelium is enhanced via adenosine Al-receptors and inhibited by adenosine A2-receptors. *J Immunol.* 1992; 148: 2201–06.

[159] Asako H, Wolf R, Granger DN. Leukocyte adherence in rat mesenteric venules: effects of adenosine and methotrexate. *Gastroenterology.* 1993; 104: 31–37.

[160] Asako H, Kubes P, Baethge BA, Wolf RE, Granger DN. Colchicine and methotrexate reduce leukocyte adherence and emigration in rat mesenteric venules. *Inflammation.* 1992; 16: 45–56.

[161] Nolte D, Lehr HA, Messmer K. Adenosinc inhibits postischemic leukocyte-endothelial interactions in postcapillary venules of the hamster. *Am J Physiol.* 1991; 261: H651–55.

[162] Cronstein BN, Eberle MA, Gruber HE, Levin RI. Methotrexate inhibits neutrophil function by stimulating adenosine release from connective tissue cells. *Proc Natl Acad Sci USA.* 1991; 88: 2441–45.

[163] Erlansson M, Bergqvist D, Persson NH, Svensjo E. Modification of post-ischemic increase of leukocyte adhesion and vascular permeability in the hamster by iloprost. *Prostaglandins.* 1991; 41: 157–68.

[164] Atherton A, Born GVR. Relationship between the velocity of rolling granulocytes and that of the blood flow in venules. *J Physiol.* 1973; 233: 157–65.

[165] Perry MA, Granger DN. Role of CD11/CD18 in shear rate-dependent leukocyte-endothelial cell interactions in cat mesenteric venules. *J Clin Invest.* 1991; 87: 1798–1804.

[166] Nazziola E, House SD. Effects of hydrodynamic and leukocyte-endothelium specificity on leukocyte-endothelium interactions. *Microvasc Res.* 1992; 44: 127–42.

[167] Bienvenu K, Russell J, Granger DN. Leukotniene B4 mediates shear rate-dependent leu-kocyte adhesion in mesenteric venules. *Circ Res*. 1992; 71: 906–11.

[168] Firrel JC, Lipowsky HH. Leukocyte margination and deformation in mesenteric venules of rat. *Am J Physiol*. 1989; 256: H1667–74.

[169] Ley K, Gaehtgens P. Endothelial, not hemodynamic, differences are responsible for prefer-ential leukocyte rolling in rat mesenteric venules. *Circ Res*. 1991; 69: 1034–41.

[170] van der Veen BS, de Winther MP, Heeringa P. Myeloperoxidase: molecular mechanisms of action and their relevance to human health and disease. *Antioxid Redox Signal*. 2009; 11: 2899–37.

[171] Lau D, Baldus S. Myeloperoxidase and its contributory role in inflammatory vascular dis-ease. *Pharmacol Ther*. 2006; 111: 16–26.

[172] Weiss SJ, Peppin G, Ortiz X, Ragsdale C, Test ST. Oxidative autoactivation of latent col-lagenase by human neutrophils. *Science*. 1985; 227: 747–49.

[173] Rowe RG, Weiss SJ. Breaching the basement membrane: who, when and how? *Trends Cell Biol*. 2008; 18: 560–74.

[174] Weiss SJ. Tissue destruction by neutrophils. *N Engl J Med*. 1989; 320: 365–76.

[175] Granger DN, Grisham MB, Kvietys PR. Mechanisms of microvascular injury. In: *Physiology of the Gastrointestinal Tract* (Johnson LR, ed.), Raven Press, New York, 1994; Chapt. 49: 1693–1722.

[176] Robinson JM. Phagocytic leukocytes and reactive oxygen species. *Histochem Cell Biol*. 2009; 131: 465–69.

[177] Cepinskas G, Noseworthy R, Kvietys PR. Transendothelial neutrophil migration. Role of neutrophil-derived proteases and relationship to transendothelial protein movement. *Circ Res*. 1997; 81: 618–26.

[178] Weiss SJ, Peppin GJ. Collagenolytic metalloenzymes of the human neutrophil. Character-istics, regulation and potential function in vivo. *Biochem Pharmacol*. 1986; 35: 3189–97.

[179] Mulivor AW, Lipowsky HH. Inhibition of glycan shedding and leukocyte-endothelial ad-hesion in postcapillary venules by suppression of matrixmetalloprotease activity with doxy-cycline. *Microcirculation*. 2009; 16: 657–66.

[180] Vink H, Constantinescu AA, Spaan JA. Oxidized lipoproteins degrade the endothelial surface layer: implications for platelet-endothelial cell adhesion. *Circulation*. 2000; 101: 1500–02.

[181] Celi A, Lorenzet R, Furie B, Furie BC. Platelet-leukocyte-endothelial cell interaction on the blood vessel wall. *Semin Hematol*. 1997; 34: 327–35.

[182] Klinger MHF. Inflammation. In: *Platelets* (Michelson AD, ed.), Elsevier Academic Press, San Diego, 2000, Chapt. 31: 459–67.

[183] Gawaz M, Langer H, May AE. Platelets in inflammation and atherogenesis. *J Clin Invest*. 2005; 115(12): 3378–84.

[184]   Tailor A, Cooper D, Granger DN. Platelet-vessel wall interactions in the microcirculation. *Microcirculation*. 2005; 12: 275–85.

[185]   Tabuchi A, Kuebler WM. Endothelium-platelet interactions in inflammatory lung disease. *Vascul Pharmacol*. 2008; 49: 141–50.

[186]   Rumbaut RE, Thiagarajan P. Platelet-vessel wall interactions in hemostasis and thrombosis. Colloquim Series in *Integrated Systems Physiology: From Molecule to Function* (Granger DN & Granger JP, eds.), Morgan-Claypool Life Sciences, 2010.

[187]   Yoshida H, Granger DN. Inflammatory bowel disease: a paradigm for the link between coagulation and inflammation. *Inflamm Bowel Dis*. 2009; 15: 1245–55.

[188]   André P, Denis CV, Ware J, Saffaripour S, Hynes RO, Ruggeri ZM, Wagner DD. Platelets adhere to and translocate on von Willebrand factor presented by endothelium in stimulated veins. *Blood*. 2000; 96: 3322–28.

[189]   Sun G, Chang WL, Li J, Berney SM, Kimpel D, van der Heyde HC. Inhibition of platelet adherence to brain microvasculature protects against severe Plasmodium berghei malaria. *Infect Immun*. 2003; 71: 6553–61.

[190]   Tailor A, Granger DN. Hypercholesterolemia promotes P-selectin-dependent platelet-endothelial cell adhesion in postcapillary venules. *Arterioscler Thromb Vasc Biol*. 2003; 23: 675–80.

[191]   Collins CE, Cahill MR, Newland AC, et al. Platelets circulate in an activated state in inflammatory bowel disease. *Gastroenterology*. 1994; 106: 840–45.

[192]   Danese S, Katz JA, Saibeni S, et al. Activated platelets are the source of elevated levels of soluble CD40 ligand in the circulation of inflammatory bowel disease patients. *Gut*. 2003; 52: 1435–41.

[193]   Collins CE, Rampton DS. Review article: platelets in inflammatory bowel disease—pathogenetic role and therapeutic implications. *Aliment Pharmacol Ther*. 1997; 11: 237–47.

[194]   Webberley MJ, Hart MT, Melikian V. Thromboembolism in inflammatory bowel disease: role of platelets. *Gut*. 1993; 34: 247–51.

[195]   Stokes KY, Calahan L, Hamric CM, Russell JM, Granger DN. CD40/CD40L contributes to hypercholesterolemia-induced microvascular inflammation. *Am J Physiol Heart Circ Physiol*. 2009; 296: H689–97.

[196]   André P, Prasad KS, Denis CV, He M, Papalia JM, Hynes RO, Phillips DR, Wagner DD. CD40L stabilizes arterial thrombi by a beta3 integrin—dependent mechanism. *Nat Med*. 2002; 8: 247–52.

[197]   Banz Y, Beldi G, Wu Y, Atkinson B, Usheva A, Robson SC. CD39 is incorporated into plasma microparticles where it maintains functional properties and impacts endothelial activation. *Br J Haematol*. 2008; 142: 627–37.

[198] Stokes KY, Russell JM, Jennings MH, Alexander JS, Granger DN. Platelet-associated NAD(P)H oxidase contributes to the thrombogenic phenotype induced by hypercholesterolemia. *Free Radic Biol Med.* 2007; 43: 22–30.

[199] Förstermann U. Nitric oxide and oxidative stress in vascular disease. *Pflugers Arch.* 2010 Mar 21. [Epub ahead of print]

[200] Arthur JF, Gardiner EE, Kenny D, Andrews RK, Berndt MC. Platelet receptor redox regulation. *Platelets.* 2008; 19: 1–8.

[201] Suzuki K, Sugimura K, Hasegawa K, et al. Activated platelets in ulcerative colitis enhance the production of reactive oxygen species by polymorphonuclear leukocytes. *Scand J Gastroenterol.* 2001; 36: 1301–06.

[202] Palmantier R, Borgeat P. Transcellular metabolism of arachidonic acid in platelets and polymorphonuclear leukocytes activated by physiological agonists: enhancement of leukotriene B4 synthesis. *Adv Exp Med Biol.* 1991; 314: 73–89.

[203] Herd CM, Page CP. Pulmonary immune cells in health and disease: platelets. *Eur Respir J.* 1994; 7: 1145–60.

[204] Mori M, Salter JW, Vowinkel T, et al. Molecular determinants of the prothrombogenic phenotype assumed by inflamed colonic venules. *Am J Physiol Gastrointest Liver Physiol.* 2005; 288: G920–26.

[205] Danese S, de la Motte C, Sturm A, et al. Platelets trigger a CD40-dependent inflammatory response in the microvasculature of inflammatory bowel disease patients. *Gastroenterology.* 2003; 124: 1249–64.

[206] Hatoum OA, Miura H, Binion DG. The vascular contribution in the pathogenesis of inflammatory bowel disease. *Am J Physiol Heart Circ Physiol.* 2003; 285: H1791–96.

[207] Guzik TJ, Hoch NE, Brown KA, et al. Role of the T cell in the genesis of angiotensin II induced hypertension and vascular dysfunction. *J Exp Med.* 2007; 204: 2449–60.

[208] Cerwinka WH, Cooper D, Krieglstein CF, Ross CR, McCord JM, Granger DN. Superoxide mediates endotoxin-induced platelet-endothelial cell adhesion in intestinal venules. *Am J Physiol Heart Circ Physiol.* 2003; 284: H535–41.

[209] Ishikawa M, Stokes KY, Zhang JH, Nanda A, Granger DN. Cerebral microvascular responses to hypercholesterolemia: roles of NADPH oxidase and P-selectin. *Circ Res.* 2004; 94: 239–44.

[210] Wood KC, Hebbel RP, Granger DN. Endothelial cell P-selectin mediates a proinflammatory and prothrombogenic phenotype in cerebral venules of sickle cell transgenic mice. *Am J Physiol Heart Circ Physiol.* 2004; 286: H1608–14.

[211] Andrews RK, Berndt MC. Platelet physiology and thrombosis. *Thrombosis Res.* 2004; 114: 447–53.

[212] Massberg S, Enders G, Leiderer R, Eisenmenger S, Vestweber D, Krombach F, Messmer K. Platelet-endothelial cell interactions during ischemia/reperfusion: the role of P-selectin. *Blood.* 1998; 92: 507–15.

[213] Massberg S, Enders G, Matos FC, Tomic LI, Leiderer R, Eisenmenger S, Messmer K, Krombach F. Fibrinogen deposition at the postischemic vessel wall promotes platelet adhesion during ischemia-reperfusion in vivo. *Blood.* 1999; 94: 3829–38.

[214] Cooper D, Chitman KD, Williams MC, Granger DN. Time-dependent platelet-vessel wall interactions induced by intestinal ischemia-reperfusion. *Am J Physiol Gastrointest Liver Physiol.* 2003; 284: G1027–33.

[215] Ishikawa M, Cooper D, Russell J, Salter JW, Zhang JH, Nanda A, Granger DN. Molecular determinants of the prothrombogenic and inflammatory phenotype assumed by the postischemic cerebral microcirculation. *Stroke.* 2003; 34: 1777–82.

[216] Cooper D, Russell J, Chitman KD, Williams MC, Wolf RE, Granger DN. Leukocyte dependence of platelet adhesion in postcapillary venules. *Am J Physiol Heart Circ Physiol.* 2004; 286: H1895–900.

[217] Vowinkel T, Wood KC, Stokes KY, Russell J, Tailor A, Anthoni C, Senninger N, Krieglstein CF, Granger DN. Mechanisms of platelet and leukocyte recruitment in experimental colitis. *Am J Physiol Gastrointest Liver Physiol.* 2007; 293: G1054–60.

[218] Tailor A, Granger DN. Hypercholesterolemia promotes leukocyte-dependent platelet adhesion in murine postcapillary venules. *Microcirculation.* 2004; 11: 597–603.

[219] Rosin C, Brunner M, Lehr S, Quehenberger P, Panzer S. The formation of platelet-leukocyte aggregates varies during the menstrual cycle. *Platelets.* 2006; 17: 61–66.

[220] Lehr HA, Messmer K. The microcirculation in atherogenesis. *Cardiovasc Res.* 1996; 32: 781–88.

[221] Lehr HA. Microcirculatory dysfunction induced by cigarette smoking. *Microcirculation.* 2000; 7: 367–84.

[222] Irving PM, Macey MG, Shah U, Webb L, Langmead L, Rampton DS. Formation of platelet-leukocyte aggregates in inflammatory bowel disease. *Inflamm Bowel Dis.* 2004; 10: 361–72.

[223] Li N, Hu H, Lindqvist M, Wikström-Jonsson E, Goodall AH, Hjemdahl P. Platelet-leukocyte cross talk in whole blood. *Arterioscler Thromb Vasc Biol.* 2000; 20: 2702–08.

[224] McGregor L, Martin J, McGregor JL. Platelet-leukocyte aggregates and derived microparticles in inflammation, vascular remodelling and thrombosis. *Front Biosci.* 2006; 11: 830–37.

[225] Rumbaut RE, Slaaf DW, Burns AR. Microvascular thrombosis models in venules and arterioles in vivo. *Microcirculation.* 2005; 12: 259–74.

[226] Egbrink MG, Van Gestel MA, Broeders MA, Tangelder GJ, Heemskerk JM, Reneman RS,

Slaaf DW. Regulation of microvascular thromboembolism in vivo. *Microcirculation*. 2005; 12: 287–300.

[227] Furie B, Furie BC. Mechanisms of thrombus formation. *N Engl J Med*. 2008; 359: 938–49.

[228] Rumbaut RE, Bellers RV, Randhawa JK, Shrimpton CN, Dasgupta S, Dong JF, Burns AR. Endotoxin enhances microvascular thrombosis in mouse cremaster muscle via a TLR-4 dependent, neutrophil-independent mechanism. *Am J Physiol*. 2006; 290: H1671–79.

[229] Broeders MA, Tangelder GJ, Slaaf DW, Reneman RS, oude Egbrink MG. Hypercholesterolemia enhances thromboembolism in arterioles but not venules: complete reversal by L-arginine. *Arterioscler Thromb Vasc Biol*. 2002; 22: 680–85.

[230] Anthoni C, Russell J, Wood KC, Stokes KY, Vowinkel T, Kirchhofer D, Granger DN: Tissue factor: a mediator of inflammatory cell recruitment, tissue injury, and thrombus formation in experimental colitis. *J Exp Med*. 2007; 204: 1595–1601.

[231] Patel KN, Soubra SH, Lam FW, Rodriguez MA, Rumbaut RE. Polymicrobial sepsis and endotoxemia promote microvascular thrombosis via distinct mechanisms. *J Thromb Haemost*. 2010 Mar 13. [Epub ahead of print]

[232] Lominadze D, Joshua IG, Schuschke DA. In vivo platelet thrombus formation in microvessels of spontaneously hypertensive rats. *Am J Hypertens*. 1997; 10: 1140–46.

[233] Esmon CT. Crosstalk between inflammation and thrombosis. *Maturitas*. 2004; 47: 305–14.

[234] Esmon CT. The interactions between inflammation and coagulation. *Br J Haematol*. 2005; 131: 417–30.

[235] Xu J, Lupu F, Esmon CT. Inflammation, innate immunity and blood coagulation. *Hamostaseologie*. 2010; 30: 5–9.

[236] Levi M, van der Poll T. Two-way interactions between inflammation and coagulation. *Trends Cardiovasc Med*. 2005; 15: 254–59.

[237] Altieri DC. Interface between inflammation and coagulation. In: *Physiology of Inflammation* (Ley K, ed.), Oxford University Press, New York, 2001; Chapt. 19: 402–22.

[238] Danese S, Dejana E, Fiocchi C. Immune regulation by microvascular endothelial cells: directing innate and adaptive immunity, coagulation, and inflammation. *J Immunol*. 2007; 178: 6017–22.

[239] Twig G, Zandman-Goddard G, Szyper-Kravitz M, Shoenfeld Y. Systemic Thromboembolism in inflammatory bowel disease. Mechanisms and clinical applications. *Ann NY Acad Sci*. 2005; 1051: 166–73.

[240] Danese S, Papa A, Saibeni S, Repici A, Malesci A, Vecchi M. Inflammation and coagulation in inflammatory bowel disease: The clot thickens. *Am J Gastroenterol*. 2007; 102: 174–86.

[241] Yoshida H, Russell J, Stokes KY, Yilmaz CE, Esmon CT, Granger DN. Role of the protein C pathway in the extraintestinal thrombosis associated with murine colitis. *Gastroenterology*. 2008; 135: 882–88.

[242] Yoshida H, Russell J, Granger DN. Thrombin mediates the extraintestinal thrombosis associated with experimental colitis. *Am J Physiol Gastrointest Liver Physiol*. 2008; 295: G904–08.

[243] Ostrovsky L, Woodman RC, Payne D, et al. Antithrombin III prevents and rapidly reverses leukocyte recruitment in ischemia/reperfusion. *Circulation*. 1997; 96: 2302–10.

[244] Sharma L, Melis E, Hickey MJ, et al. The cytoplasmic domain of tissue factor contributes to leukocyte recruitment and death in endotoxemia. *Am J Pathol*. 2004; 165: 331–40.

[245] Arnold CS, Parker C, Upshaw R, et al. The antithrombotic and anti-inflammatory effects of BCX-3607, a small molecule tissue factor/factor VIIa inhibitor. *Thromb Res*. 2006; 117: 343–49.

[246] Levi M, Dörffler-Melly J, Reitsma P, et al. Aggravation of endotoxin-induced disseminated intravascular coagulation and cytokine activation in heterozygous protein-C-deficient mice. *Blood*. 2003; 101: 4823–27.

[247] Rumbaut RE, Randhawa JK, Smith CW, Burns AR. Mouse cremaster venules are predisposed to light/dye-induced thrombosis independent of wall shear rate, CD18, ICAM-1, or P-selectin. *Microcirculation*. 2004; 11: 239–47.

[248] Senchenkova EY, Granger DN. T-lymphocytes contribute to angiotensin II-mediated thrombosis in cremaster muscle arterioles. *FASEB J*. 2010; 24: 589.3.

[249] Woldhuis B, Tangelder GJ, Slaaf DW, Reneman RS. Concentration profile of blood platelets in arterioles and venules. *Am J Physiol*. 1992; H1217–23.

[250] Kuijpers MJ, Munnix IC, Cosemans JM, Vlijmen BV, Reutelingsperger CP, Egbrink MO, Heemskerk JW. Key role of platelet procoagulant activity in tissue factor-and collagen-dependent thrombus formation in arterioles and venules in vivo differential sensitivity to thrombin inhibition. *Microcirculation*. 2008; 15: 269–82.

[251] Clark, SR, et al. Platelet TLR4 activates neutrophil extracellular traps to ensnare bacteria in septic blood. *Nat Med*. 2007; 13: 463–69.

[252] Hidalgo, A, et al. Heterotypic interactions enabled by polarized neutrophil microdomains mediate thromboinflammatory injury. *Nat Med*. 2009; 15: 384–91.

[253] Ma AC, Kubes P. Platelets, neutrophils, and neutrophil extracellular traps (NETs) in sepsis. *J Thromb Haemost*. 2008; 6: 415–20.

[254] Nawroth PP, Stern DM. Modulation of endothelial cell hemostatic properties by tumor necrosis factor. *J Exp Med*. 1986; 163: 740–45.

[255]   Szotowski B, Antoniak S, Poller W, Schultheiss HP, Rauch U. Procoagulant soluble tissue factor is released from endothelial cells in response to inflammatory cytokines. *Circ Res.* 2005; 96: 1233–39.

[256]   Takano S, Kimura S, Ohdama S, Aoki N. Plasma thrombomodulin in health and diseases. *Blood.* 1990; 76: 2024–29.

[257]   Robson SC, Sévigny J, Zimmermann H. The E-NTPDase family of ectonucleotidases: Structure function relationships and pathophysiological significance. *Purinergic Signal.* 2006; 2: 409–30.

[258]   Dwyer KM, Deaglio S, Gao W, Friedman D, Strom TB, Robson SC. CD39 and control of cellular immune responses. *Purinergic Signal.* 2007; 3: 171–80.

[259]   Dwyer KM, Robson SC, Nandurkar HH, Campbell DJ, Gock H, Murray-Segal LJ, Fisicaro N, Mysore TB, Kaczmarek E, Cowan PJ, d'Apice AJ. Thromboregulatory manifestations in human CD39 transgenic mice and the implications for thrombotic disease and transplantation. *J Clin Invest.* 2004; 113: 1440–46.

[260]   Vowinkel T, Anthoni C, Wood KC, Stokes KY, Russell J, Gray L, Bharwani S, Senninger N, Alexander JS, Krieglstein CF, Grisham MB, Granger DN. CD40-CD40 ligand mediates the recruitment of leukocytes and platelets in the inflamed murine colon. *Gastroenterology.* 2007; 132: 955–65.

[261]   Antoniades C, Bakogiannis C, Tousoulis D, Antonopoulos AS, Stefanadis C. The CD40/CD40 ligand system: linking inflammation with atherothrombosis. *J Am Coll Cardiol.* 2009; 54: 669–77.

[262]   Freedman JE. Oxidative stress and platelets. *Arterioscler Thromb Vasc Biol.* 2008; 28: s11–s16.

[263]   Rosenblum WI, El Sabban F. Dimethylsulfoxide and glycerol, hydroxyl radical scavengers, impair platelet aggregation within and eliminate the accompanying vasodilation of, injured mouse pial arterioles. *Stroke.* 1982; 13: 35–39.

[264]   Peire MA, Puig-Parallada P. Oxygen free radicals and nitric oxide are involved in the thrombus growth produced by iontophoresis of ADP. *Pharmacol Res.* 1998; 38: 353–56.

[265]   Lindberg RA, Slaaf DW, Lentsch AB, Miller FN. Involvement of nitric oxide and cyclooxygenase products in photoactivation-induced microvascular occlusion. *Microvasc Res.* 1994; 47: 203–21.

[266]   Sasaki Y, Seki J, Giddings JC, Yamamoto J. Effects of NO-donors on thrombus formation and microcirculation in cerebral vessels of the rat. *Thromb Haemostasis.* 1996; 76: 111–17.

[267]   Broeders MA, Tangelder GJ, Slaaf DW, Reneman RS, oude Egbrink MG. Endogenous nitric oxide protects against thromboembolism in venules but not in arterioles. *Arterioscler Thromb Vasc Biol.* 1998; 18: 139–45.

[268]   Kumar P, Shen Q, Pivetti CD, Lee ES, Wu MH, Yuan SY. Molecular mechanisms of endothelial hypermeability: implications in inflammation. *Expert Rev Mol Med.* 2009; Jun 30; 11: e19.

[269]   Weber C, Fraemohs L, Dejana E. The role of junctional adhesion molecules in vascular inflammation. *Nat Rev Immunol.* 2007; 7: 467–77.

[270]   Aghajanian A, Wittchen ES, Allingham MJ, Garrett TA, Burridge K. Endothelial cell junctions and the regulation of vascular permeability and leukocyte transmigration. *J Thromb Haemost* 2008; 6: 1453–60.

[271]   Dvorak AM, Feng D. The vesiculo-vacuolar organelle (VVO). A new endothelial cell permeability organelle. *J Histochem Cytochem.* 2001; 49: 419–32.

[272]   Nagy J, Benjamin L, Zeng H, Dvorak AM, Dvorak HF. Vascular permeability, vascular hyperpermeability and angiogenesis. *Angiogenesis.* 2008; 11: 109–19.

[273]   Bogatcheva NV, Verin AD. The role of the cytoskeleton in the regulation of vascular endothelial barrier function. *Microvasc Res.* 2008; 76: 202–27.

[274]   Pilati CF. Macromolecular transport in canine coronary microvasculature. *Am J Physiol.* 1990; 258: H748–53.

[275]   Victorino GP, Ramirez RM, Chong TJ, Curran B, Sadjadi J. Ischemia-reperfusion injury in rats affects hydraulic conductivity in two phases that are temporally and mechanistically separate. *Am J Physiol Heart Circ Physiol.* 2008; 295: H2164–71.

[276]   Oliver MG, Specian RD, Perry MA, Granger DN. Morphologic assessment of leukocyte-endothelial cell interactions in mesenteric venules subjected to ischemia and reperfusion. *Inflammation.* 1991; 15: 331–46.

[277]   Yang Z, Sharma AK, Linden J, Kron IL, Laubach VE. CD4+ T lymphocytes mediate acute pulmonary ischemia-reperfusion injury. *J Thorac Cardiovasc Surg.* 2009; 137: 695–702.

[278]   Khandoga AG, Khandoga A, Anders HJ, Krombach F. Postischemic vascular permeability requires both TLR-2 and TLR-4, but only TLR-2 mediates the transendothelial migration of leukocytes. *Shock.* 2009; 31: 592–98.

[279]   Hernandez LA, Grisham MB, Twohig B, Arfors KE, Harlan JM, Granger DN. Role of neutrophils in ischemia-reperfusion-induced microvascular injury. *Am J Physiol.* 1987; 253: H699–703.

[280]   Kurose I, Wolf R, Grisham MB, Granger DN. Modulation of ischemia/reperfusion-induced microvascular dysfunction by nitric oxide. *Circ Res.* 1994; 74: 376–82.

[281]   Nagai M, Terao S, Yilmaz G, Yilmaz CE, Esmon CT, Watanabe E et al. Roles of inflammation and the activated protein C pathway in the brain edema associated with cerebral venous sinus thrombosis. *Stroke.* 2009. PMID: 19892996.

[282] Eppinger MJ, Jones ML, Deeb GM, Bolling SF, Ward PA. Pattern of injury and the role of neutrophils in reperfusion injury of rat lung. *J Surg Res.* 1995; 58: 713–18.

[283] Carden DL, Smith JK, Korthuis RJ. Neutrophil-mediated microvascular dysfunction in post-ischemic canine skeletal muscle. Role of granulocyte adherence. *Circ Res.* 1990; 66: 1436–44.

[284] Levine AJ, Parkes K, Rooney SJ, Bonser RS. The effect of adhesion molecule blockade on pulmonary reperfusion injury. *Ann Thorac Surg.* 2002; 73: 1101–06.

[285] Kluger MS. Vascular endothelial cell adhesion and signaling during leukocyte recruitment. *Adv Dermatol.* 2004; 20: 163–201.

[286] Ionescu CV, Cepinskas G, Savickiene J, Sandig M, Kvietys PR. Neutrophils induce sequential focal changes in endothelial adherens junction components: role of elastase. *Microcirculation.* 2003; 10: 205–20.

[287] Shigematsu T, Wolf RE, Granger DN. T-lymphocytes modulate the microvascular and inflammatory responses to intestinal ischemia-reperfusion. *Microcirculation.* 2002; 9: 99–109.

[288] Yang Z, Sharma AK, Linden J, Kron IL, Laubach VE. CD4+ T lymphocytes mediate acute pulmonary ischemia-reperfusion injury. *J Thorac Cardiovasc Surg.* 2009; 137: 695–702.

[289] Liu M, Chien CC, Grigoryev DN, Gandolfo MT, Colvin RB, Rabb H. Effect of T cells on vascular permeability in early ischemic acute kidney injury in mice. *Microvasc Res.* 2009; 77: 340–47.

[290] Reynolds JM, McDonagh PF. Platelets do not modulate leukocyte-mediated coronary microvascular damage during early reperfusion. *Am J Physiol.* 1994; 266: H171–81.

[291] Heindl B, Zahler S, Welsch U, Becker BF. Disparate effects of adhesion and degranulation of platelets on myocardial and coronary function in postischaemic hearts. *Cardiovasc Res.* 1998; 38: 383–94.

[292] Zamora CA, Baron D, Heffner JE. Washed human platelets prevent ischemia-reperfusion edema in isolated rabbit lungs. *J Appl Physiol.* 1991; 70: 1075–84.

[293] Alexander JS, Patton WF, Christman BW, Haselton FR. Platelet-derived lysophosphatidic acid decreases endothelial permeability in vitro. *Am J Physiol.* 1998; 272: H115–22.

[294] Schaphorst KL, Chiang E, Jacobs KN, Zaiman A, Natarajan V, Wigley F, Garcia JG. Role of sphingosine-1 phosphate in the enhancement of endothelial barrier integrity by platelet-released products. *Am J Physiol.* 2003; L258–67.

[295] Strbian D, Karjalainen-Lindsberg ML, Tatlisumak T, Lindsberg PJ. Cerebral mast cells regulate early ischemic brain swelling and neutrophil accumulation. *J Cereb Blood Flow Metab.* 2006; 26: 605–12.

[296] Lee KS, Kim SR, Min KH, Lee KY, Choe YH, Park SY, Chai OH, Zhang X, Song SH, Lee LC. Mast cells can mediate vascular permeability through regulation of the PI3K-HIF-1alpha-VEGF axis. *Am J Respir Crit Care Med.* 2008; 178: 787–97.

[297] Iuvone T, Den Bossche RV, D'Acquisto F, Carnuccio R, Herman AG. Evidence that mast cell degranulation, histamine and tumor necrosis factor release occur in LPS-induced plasma leakage in rat skin. *Br J Pharmacol.* 1999; 128: 700–04.

[298] Miller HR, Pemberton AD. Tissue-specific expression of mast cell granule serine proteinases and their role in inflammation in the lung and gut. *Immunology.* 2002; 105: 375–90.

[299] Naidu BV, Krishnadasan B, Farivar AS, Woolley SM, Thomas R, Van Rooijen N, et al. Early activation of the alveolar macrophage is critical to the development of lung ischemia-reperfusion injury. *J Thorac Cardiovasc Surg.* 2003; 126: 200–07.

[300] Chen Y, Lui VC, Rooijen NV, Tam PK. Depletion of intestinal resident macrophages prevents ischaemia reperfusion injury in gut. *Gut.* 2004; 53: 1772–80.

[301] Zhao M, Fernandez LG, Doctor A, Sharma AK, Zarbock A, Tribble CG, et al. Alveolar macrophage activation is a key initiation signal for acute lung ischemia-reperfusion injury. *Am J Physiol Lung Cell Mol Physiol.* 2006; 291: L1018–26.

[302] Kubota Y, Iwasaki Y, Harada H, Yokomura I, Ueda M, Hashimoto S, Nakagawa M. Depletion of alveolar macrophages by treatment with 2-chloroadenosine aerosol. *Clin Diagn Lab Immunol.* 1999; 6: 452–56.

[303] Gray M, Palispis W, Popovich PG, van Rooijen N, Gupta R. Macrophage depletion alters the blood-nerve barrier without affecting Schwann cell function after neural injury. *J Neurosci Res.* 2007; 85: 766–77.

[304] Inauen W, Payne DK, Kvietys PR, Granger DN. Hypoxia/reoxygenation increases the permeability of endothelial cell monolayers: role of oxygen radicals. *Free Radic Biol Med.* 1990; 9: 219–23.

[305] Granger DN, Rutili G, McCord JM. Superoxide radicals in feline intestinal ischemia. *Gastroenterology.* 1981; 81: 22–29.

[306] Yang Z, Sharma AK, Marshall M, Kron IL, Laubach VE. NADPH oxidase in bone marrow-derived cells mediates pulmonary ischemia-reperfusion injury. *Am J Respir Cell Mol Biol.* 2009; 40: 375–81.

[307] Kahles T, Luedike P, Endres M, Galla HJ, Steinmetz H, Busse R, et al. NADPH oxidase plays a central role in blood-brain barrier damage in experimental stroke. *Stroke.* 2007; 38: 3000–06.

[308] Gertzberg N, Neumann P, Rizzo V, Johnson A. NAD(P)H oxidase mediates the endothelial barrier dysfunction induced by TNF-alpha. *Am J Physiol.* 2004; L37–48.

[309] Warboys CM, Toh HB, Fraser PA. Role of NADPH oxidase in retinal microvascular permeability increase by RAGE activation. *Invest Ophthalmol Vis Sci.* 2009; 50: 1319–28.

[310] Kurose I, Wolf R, Grisham MB, Granger DN. Modulation of ischemia/reperfusion-induced microvascular dysfunction by nitric oxide. *Circ Res.* 1994; 74: 376–82.

[311] Ozaki M, Kawashima S, Hirase T, Yamashita T, Namiki M, Inoue N, et al. Overexpression of endothelial nitric oxide synthase in endothelial cells is protective against ischemia-reperfusion injury in mouse skeletal muscle. *Am J Pathol.* 2002; 160: 1335–44.

[312] Kurose I, Kubes P, Wolf R, Anderson DC, Paulson J, Miyasaka M, et al. Inhibition of nitric oxide production. Mechanisms of vascular albumin leakage. *Circ Res.* 1993; 73: 164–71.

[313] Rumbaut RE, Wang J, Huxley VH. Differential effects of L-NAME on rat venular hydraulic conductivity. *Am J Physiol Heart Circ Physiol.* 2000; 279: H2017–23.

[314] Hiratsuka M, Katayama T, Uematsu K, Kiyomura M, Ito M. In vivo visualization of nitric oxide and interactions among platelets, leukocytes, and endothelium following hemorrhagic shock and reperfusion. *Inflamm Res.* 2009; 58: 463–71.

[315] Vachharajani V, Granger DN. Adipose tissue: a motor for the inflammation associated with obesity. *IUBMB Life.* 2009; 61: 424–30.

[316] Rodrigues SF, Granger DN. Role of blood cells in ischaemia-reperfusion induced endothelial barrier failure. *Cardiovasc Res.* 2010 Apr 13. [Epub ahead of print]